Y0-BDO-085

5/20/88 Edward R. Hamilton 58/22 3.95

. . . what science really is — not a trivial business of tricky hardware, not the phony bubbling retorts of the advertisements, not the strange men with white coats or beards, but the response, at once poetic and analytical, of man's creative mind to the challenge of the mystery of matter and life.

William Weaver in
GOALS FOR AMERICANS

SCIENCE, TECHNOLOG

ND HUMAN VALUES

A. Cornelius Benjamin

UNIVERSITY OF MISSOURI PRESS • COLUMBIA

ISBN 0-8262-0035-4

University of Missouri Press, Columbia, Missouri 65201
Library of Congress Catalog Card Number 65-10698
Printed and bound in the United States of America

Reprinted 1974, 1976

Acknowledgements

The ideas presented in the following pages have been drawn from such a wide variety of areas—scientific, philosophical, and literary—and have developed through so many years of teaching and research that I am unable to trace their origins. Perhaps it is to my students, whose persistent and often penetrating questions have sharpened and broadened my thinking, that I owe my greatest debt of gratitude. I am very grateful to the staff of the University of Missouri Press for valuable editorial aid, so generously given. I wish to thank the University Research Council for financial assistance in preparing the manuscript. I am especially indebted to my wife, who helped with much of the work involved in putting the material into publishable form, and who assisted in preparing the index.

Contents

Introduction

Man is a creature of hope, but he is also a creature of action. In times of despair, when the burdens of the world have rested heavily on his shoulders, he has not in general—at least in the Western world—found relief in blind resignation. On the contrary he has sought help by taking and using whatever devices his own culture provided. Frequently these were nothing better than magic, superstition, astrology, and other forms of pseudo science. Sometimes they were religion and worship, with the propitiatory tools of prayer and sacrifice. Often they were elaborate philosophies and other intellectual constructions which endeavored to soften the evil of the world either by boldly calling it unreal, or by arguing that it was something which must be suffered and endured because only by its presence could the good, by contrast, really *be* good.

None of these tools has been universally successful, and man, in his persistent search for happiness, has sought more effective means. Very early in his development he discovered an instrument which seemed to offer great promise. This was science. None of the limitations inherent in the other devices seemed to be found in science. Since its powers were natural rather than supernatural it could not properly refuse to cooperate with man—as religion frequently did—on the grounds that the demands which were being put upon it were unrighteous. Furthermore, in order to use science man did not have to meet the very exacting but somewhat unreasonable conditions for practicing the black art. Nor was it as unreliable as superstition or as unproductive of hoped-for results as scanning the heavens. On the positive side science seemed to eliminate evil from the world, hence faced the problem squarely and did not compel man to practice the systematic deceit which was involved in all attempts to make evil less evil by calling it "good." True, science had one limitation—it was still a mere youth. But growth seemed inevitable, and man readily concluded that with the passing years, science would provide the instrument on which he could truly rely for the attainment of final and complete happiness.

The progress of science through the years has done much to confirm this hope. Man is more comfortable than he has ever been before. Drudgery, extreme exertion, long hours of labor, excessive heat and cold, poor lighting, unnecessary noise—all have been reduced to a minimum. Science has greatly reduced suffering and pain, is progressively gaining control over disease, and, while it has not eliminated the fact of death, has to a great extent removed its terror. Man has incredibly more opportunities for cultural enjoyment than his ancestors ever dreamed of having. Through radio, television, motion pictures, stereophonic recordings, color photography, and paperbacks modern man has at his hand the learning and art of the ages.

But the discerning man of today, while he recognizes the immense contributions of science, notes with dismay that all is not well. Though it is true that he is more comfortable today, he is not satisfied with his life in contemporary society and the world. Is it possible that he is beginning to realize that mere comfort cannot produce happiness? The obvious products of science are machines and gadgets, things which are tangible, *goods* which have market value. Suppose man acquires a great abundance of such instruments of comfort; most of his physical needs will be met and he will have unlimited leisure at his disposal. What shall he then do with this leisure? Produce more material goods? But the saturation point in this area cannot be far off, in spite of the ingenuity of advertisers in attempting to create ever new wants in the mind of the buying public. Man's predicament today lies partly in his failure to realize that genuinely happy living does not lie in the search for comfort and leisure, but in the pursuit of "higher" values—cultural, humanistic, aesthetic, religious, and intellectual. These values, made available largely through science, are here for man's taking; but he has not taken them.

Furthermore, science has multiplied not only the positive values but the negative ones as well. Television has presented great dramas and informative educational programs; but it has also produced soap operas and irritating commercials. We now have accurate recordings of the great masterpieces of music; but we also have hi-fi reproductions of tunes which ought never to have been recorded in the first place. Paperbacks provide us with the classics of the written word; but they also make available unlimited copies of literary trash. It is true that man can be, if he wishes, reasonably free from pain and disease. Nevertheless, as science has multiplied the methods of cure and control, it has increased at an even greater rate the devices which produce these unhappy states. Of what value is a preventive serum for polio when a guided missile can wipe out the human race? Indeed, even when we move into the area of *pure* science we realize that strange things are happening. Science is beginning to capitalize on its well-earned popularity by claiming the right to penetrate into the most private aspects of our lives. Nothing, not even religion, morality, or aesthetic enjoyment, is spared intellectual scrutiny. Appreciation is being destroyed by critical analysis, and we run the risk of losing some of our most sacred values because science has torn them into conceptual shreds. Intellectualism has become the dominant philosophy of our times.

The plain fact of the matter seems to be, therefore, that science is not an unmixed blessing. Like most of the things in life, it has powers for good and powers for evil, depending on how it is put to work. This immediately suggests an important question: Who is to decide how science is to be used? Is it to be the scientists, who presumably know what science is, but who, just because they are practicing specialists, may not be acutely aware of the social, political, moral, and religious problems facing us? Or is it to be the social workers, the politicians, the moral and religious teachers, or even the "common man," who may not know much about science but who do recognize some of the social predicaments into which science has driven us today.

A recent writer has made a pointed observation. In ancient Greece, he says, where Democracy was the dominating social concept, people were divided into two classes: those who were actively involved in the Democratic process and those who were not. The former group held office or could hold office, while the latter were either disinclined to participate or incapable of doing so. These were called *idiotes*, from which the modern term *idiots* has come. But for the Greeks the word carried none of the connotation of mental deficiency which it implies today, and meant simply a group of people who were ignorant in a certain specified area of knowledge. The writer goes on to suggest that Science plays in our civilization today much the same role that Democracy did among the Greeks. And in much the same way we have created a social class characterized by lack of participation in, or appreciation for, this tremendous social force which is embodied in contemporary science in all of its forms. These people could also be called *idiotes*.

The Spanish philosopher, Ortega y Gasset, has another suggestion. Many scientists, he argues, in spite of being very well informed in their fields of special interest are more or less ignorant in areas of the humanities and of social and political problems. He suggests that these men be called *learned ignoramuses*, not only because they both know and do not know but because they frequently suffer from the "petulance of those who are learned"—they feel that their competence in science makes them authoritative in areas having to do with the *uses* of science and with other human values.

The choice of the term *ignoramus* here is very unfortunate. It is a derogatory word and suggests that many scientists are ignorant in an area where they *ought* to be informed. But why should they be? Certainly it is not in their role as scientists that they must know all of the implications of science for society at large, but in their role as citizens of the nation and the world. If one insists on using this emotive word he should recognize that the learned ignoramuses are not restricted to scientists. Many humanists do not know the content and method of science; yet they suffer from the same petulance, and proceed to blame science for doing things which a rational understanding of science would show not to be within its province at all. Someone recently distinguished humanists from scientists by saying that humanists are ignorant of science and ashamed of it, but scientists are ignorant of the humanities and proud of it. This is one of those convenient phrases which can be adapted to the particular audience to which one wishes to appeal. The fact seems to be that there are just as many humanists who are proud of their ignorance of science as there are scientists who are ashamed of their ignorance of the humanities. Anyone should be ashamed of his ignorance in any area in which he is expected to make rational and authoritative judgments.

The issue cannot be resolved so long as it is formulated in emotionally charged words. C. P. Snow, whose book, *The Two Cultures and the Scientific Revolution*,[1] has had an amazing popularity in recent years, and who is himself both a scientist and a novelist, proposes that we speak henceforth of the "two cultures," and thus avoid the emotional associations involved in such terms as *uncultured, unintellectual, inhumane, unscientific.* While admitting the present tragedy of humanists who are totally ignorant of science, and scientists who are totally ignorant of the humanities, he proposes that we can bring them together only by recognizing that each has something to contribute to our civilization, and that the need for cooperative effort between the two groups in the years which are immediately ahead is imperative if we are to survive.

I shall therefore try to employ neutral terms and state precisely the problem I propose to discuss. All of us know that it is sometimes difficult to see the woods for the trees. For scientists this is particularly true. Science is a severe taskmaster and allows little time and little energy for activities which are, at best, on the fringe of science and, at worst, in no significant way related to science. The scientist must study science, not its form or its environment or its social and political implications. He must first *know* science, and only after he has done this can he, in the for-

[1] (New York: Cambridge University Press, 1961.) Snow has recently published a second book, *The Two Cultures: and a Second Look* (New York: Cambridge University Press, 1964), which contains a reprint of the original book and a long additional essay restating his position and replying to some of the criticisms of it which have appeared in print and in his private correspondence. The latter book appeared too late for me to comment in the present manuscript on the ideas which it contains. Suffice it to say that he has now broadened his point of view and his ideas are more or less in accord with many of the points which I shall develop in the following pages.

tunate phrase of Dr. Conant, *understand* science. He must know the trees before he knows the woods. And often, just because he knows the trees so well, he cannot know the woods.

I think no one can deny that we live in the woods of science today. To say that we live in a scientific society is to utter a mere commonplace. And I think it is clear that, whether we want to or not, we *should* know where we are living. All of us *live* in the woods, but most of us *see* only the trees or *see* neither the woods nor the trees. This has unfortunate consequences. If we do not know the woods we may be quite unaware of what they are doing to us, and what we are capable of doing to them. The ensuing pages may be taken as a handbook in woodsmanship—a manual which may help us to live in a society which is heavily wooded by science.

My design in writing this book has been to communicate with those who are seeking this kind of knowledge about science. I have not assumed any specific familiarity with the content of any of the sciences or with any of the controversies concerning the nature and role of science. Furthermore, in carrying out my purpose I have tried to avoid technical language. Some of the material which follows I have previously published in professional journals. All of this has been completely rewritten in an attempt to make it more intelligible to the layman. This has resulted in what may appear to be a certain looseness in the language I have used, an oversimplification in some of the statements I have made, and perhaps an unjustifiable dogmatism in my approach. My only reply to this possible criticism is that if we are to direct anybody where to go we must start from where he is, and then see that we do not lose him on the way. Science and the logic of science have in recent years developed highly technical vocabularies which enable the specialists in their fields to talk successfully with one another. But when the technicians use these vocabularies they do not communicate with the man in the street, and they are therefore unable to lead him to the promised land. On the other hand, if they talk merely in the vocabulary of the layman, they find that there is really no place where they can lead him. I see no way to escape this difficulty except by adopting a language with the highest degree of precision which is compatible with the intelligence of my intended audience. I believe, with Huxley, that the method of science is simply that of common sense thinking, after this has been subjected to certain refinements and controls. I further believe that the most illuminating examples of scientific thinking are drawn from its ABC, not from its XYZ. As Bertrand Russell has suggested, we can learn about mathematics by starting with the fact that two plus two equals four, which at least *seems* perfectly clear to all of us, and then go either "down into lower mathematics" and attempt to prove that two plus two really *does* equal four, or "up into higher mathematics" and develop algebra, analytic geometry, and calculus out of such simple truths. The point is that we must know where to start and how to proceed.

In support of both my starting point and my procedure I can only call attention to the following analogy. We all believe that lying is wrong. But when we teach this truth to our children, we do not begin by specifying the many and varied exceptions which limit the generality of this moral principle; we do not trouble ourselves at the moment to inform them that there are occasions when lying is not wrong, and that there are cultures where dishonesty is socially approved. We feel sure that our children will learn these things in the course of their broadening experience. We know, with Pope, that a little learning is a dangerous thing, but we also know that a little learning is better than complete ignorance and is a necessary prerequisite to that critical kind of *un*learning which we associate with the revising and correcting of what we have already learned. I have proceeded on the assumption that my reader wishes to be informed, and that he cannot learn anything unless he is taught in a language which is reasonably easy for him to understand. I assume also that he will accept this knowledge in the way in which he tentatively accepts moral principles—not as final truths but as guiding principles subject to correction by later and more critical experience. I can only hope that in my choice of a somewhat popular vocabulary I do not merely elaborate on something which the reader already knows perfectly well, or that I do not misinform him by the inexactness of my terminology. But if I am to avoid the technical and somewhat forbidding language of the specialist, these are the risks which I must take.

1

What Is Science?

When man first began to wonder about the world, science was born. What this wonderment was like we have no way of knowing. Probably it was a crude mixture of fear, religious adoration, and pure curiosity. Whatever its nature, it was the ancestor not only of science but of poetry, religion, and philosophy as well. Even in our times the distinctions between these ways of reacting to the world are far from sharp; much of science is philosophy, much of religion is poetry. Perhaps they cannot and should not be separated from one another. Since wonder, their common parent, is a combination of the intellectual urge to understand, the practical need to control, and the aesthetic impulse to enjoy and appreciate, there is no reason why we should expect to find any one of these types of experience in its purity. The world in which man finds himself is at once terrifying and beautiful, the source of all pain and the origin of everything that is good, something which piques his curiosity yet baffles his understanding, the place in which he must both live and die. Should we be surprised, therefore, that he looks upon it with opposed and mixed emotions?

But even though science may contain elements of poetry, religion, and philosophy, in its aim and intent it is none of these things. At least if we are to understand what science really is we must eliminate these ingredients and try to discover science in its purity. We should attempt to determine how much of this complex response of man to the world, which we call "science" in the general sense of this word, is truly an expression of the urge to explain and comprehend, and how much is the result of man's attempt to find an emotional kinship with his surroundings. Even if the boundaries between science and these other experiences are unclear in actual life, they should not be in our understanding; we can properly grasp the role of science in the modern world only if we are not led astray at the very beginning by a confusion in terms.

1. Difficulties of Defining *Science*

Consequently, our first task is one of definition. What is science?[1] Immediately we run up against two difficulties. In the first place, all definition is to a certain extent arbitrary. We can, like Humpty-Dumpty, make words mean exactly what we choose to have them mean. Strictly speaking we are free as the wind in assigning any noise we please to any kind of object. When scientists are called upon to invent words to designate newly discovered objects—electrons, protons, genes, chromosomes—they do just this, except that they call on the classical scholars to give their names a certain dignity. If we wished we could select anything whatsoever from our experience and decide, quite arbitrarily, to call it "science." On grounds of logic no one could possibly charge us with having made a mistake, any more than he could convict us of error if we decided to call our new son "Charles" rather than "Henry."

The reason we do not do this, of course, is that the assigning of common names is not precisely the same thing as the assigning of proper names. Common names become the property of society and are instruments of communication. This fact puts us under certain obligations to abide by social usage. When Eve saw Adam for the first time, as Mark Twain tells us in *Eve's Diary*,

[1] Most people do not realize how recently the word "science" was added to the English language. It was first used by William Whewell in 1840, but it was not included in the *Oxford English Dictionary* until 1867.

she decided to call him a "man." Why? Because he *looked* like a man. If we are to use the word "science" we ought to apply it to something that *looks* like science. The word has had a long history and now has a dictionary definition, and we ought, for the purposes of making ourselves understood by other people, to abide by this convention.

The second difficulty in arriving at a good definition of "science" is much more serious. The word has become a term of praise: the scientific way of doing things is the "best" way. The scientist is a member of the social élite. Scientific detective stories are better than all others; advertised articles which bear the stamp of scientific approval sell best;[2] a scientifically controlled and administered society, though perhaps still far in the future, would be Utopia itself; any study, such as sociology or economics, reaches intellectual maturity only when it adopts the techniques and methods of science; and many cultural studies, fine arts, and humanities would welcome the dignity which they should acquire if they could truthfully be called "sciences." Even a religion which is scientific appears to many people to be a better kind of religion. Thus the emotional appeal of the word has led to its widespread use in situations which are quite foreign to its proper meaning. It has become a dangerous word—dangerous in the sense of leading to confusion in thinking and producing misleading associations. When anything calls itself "science," beware!

In the face of these difficulties what should be our procedure in attempting to define "science"? We wish to use the word as far as possible in conformity with actual practice. Otherwise we run the risk of being completely misunderstood when we employ it. But we must acknowledge frankly that custom admits of a variety of uses all of which cannot, in consistency, be included in our definition. A good dictionary gives five or six definitions of the term. At this point, therefore, a more or less arbitrary decision enters. All that we can do is to bring together the studies which are commonly called by this name, realizing that there will be many borderline cases which are called "sciences" by certain people, and not by others. The use of the common suffixes *-ology, -osophy, -onomy,* and *-ography* to distinguish the genuine sciences from the

---

[2] Weaver notes that "research" with a "smellometer" is used by advertisers to indicate the supposed quantitative effectiveness of a deodorant. Warren Weaver, "A Great Age for Science," *Goals for Americans: The Report of the President's Commission on National Goals* (Englewood Cliffs, New Jersey: Prentice-Hall, Inc., 1960), p. 118.

pseudo sciences cannot be fruitful, since such charlatans as astrologists, graphologists, and phrenologists adopt them quite as readily as do such genuine scientists as biologists and geologists. In other words, we must be prepared for the fact that by defining "science" in this loose, empirical way we are bound to admit many impostors whom we shall later have to reject. But with the motley array before us we can attempt to extract what seem to be the common features. If they are all sciences they presumably are in some respects alike.

However, the problem is not nearly so simple. Definitions can be unsatisfactory because they are too broad or because they are too narrow. Suppose, for example, we are trying to define the concept "human being," by finding as many and as varied instances as we can. This will certainly cause difficulties. Shall we include a man who is hopelessly insane or a man in a drunken rage? Shall we include an imbecile who cannot speak and whose every want must be taken care of by someone else? Shall we include a three-month premature babe living in an incubator, or a person dying in a coma, or a child born without arms or legs? The feature common to these creatures and to normal human beings would be insufficiently restrictive; our definition would cover all men only in the sense of their being living organisms, and would therefore be much too broad. On the other hand, if the instances we happen to start with are white men, or men with sandy hair, or men who believe in Christianity, or men who admire Beethoven, our resulting definition will be much too restrictive and narrow.

Similar considerations are involved in the attempt to define "science." We might select certain properties which, while possessed by all sciences, are common to many activities which are not science at all. A definition constructed on this basis would not be very satisfactory because it would be too broad. For example, science, religion, poetry, and philosophy, as we saw at the beginning of our discussion, are all modes of response to the world; but if we should define "science" simply as a mode of response to the world, we should be giving the term a meaning which is not sufficiently restrictive in character. On the other hand, if we were to select as the defining property of "science" a certain attribute —for example, the use of mathematical methods, possessed only by certain of the studies which are commonly called by this name—we should produce a definition which would be too narrow to be very useful.

The problem of defining "science" in this tentative manner is not a difficult one. All sciences which are commonly described by this term normally investigate or inquire into a certain subject matter for the purpose of arriving at something which is, or at least claims to be, knowledge. Every science, therefore, has five aspects.

It has, first, a *subject matter* or field over which the investigation extends. This is commonly called "nature." There will be certain advantages at the present stage of our study if we leave this word somewhat vague in meaning.[3] We do not wish, for example, to exclude the possibility of social sciences, which do not study nature in the strict sense of the word, or such a science as theology, which might presumably be defined as the science of God, or astrology, which claims to investigate the effects of the stars on human destiny. In view of this difficulty let us suppose, for the present, that while science must have a subject matter—something which it studies, something about which our knowledge is built up and to which our knowledge refers when it claims to be *genuine* knowledge—there are no restrictions whatsoever as to *what* this may be.

In the second place, every science employs a *method of procedure* by which it hopes to arrive at knowledge. This is commonly called the "experimental method," but we shall find this term much too restricted in meaning to suit our purposes. There are certainly many studies which rightfully claim to be sciences but do not employ laboratory techniques. We prefer a somewhat more sweeping term for the characterization of the scientific method—a word or phrase which will describe the method of knowing that is used by science yet does not commit us to a particular procedure which may be employed only by certain of the sciences. It seems advisable at this stage to select a term which is probably too general to serve our final purpose. Let us call it

---

[3] Attempts are commonly made at the outset of a study of science to restrict its subject matter more than is really necessary. For example, Campbell insists that science is concerned only with those judgments on which universal agreement is possible. Norman Campbell, *What is Science?* (New York: Dover Publications, 1921), Chap. II. Feigl states that scientific subject matter should be "intersubjectively testable." Herbert Feigl, "The Scientific Outlook," *Readings in the Philosophy of Science,* ed. H. Feigl and M. Brodbeck (New York: Appleton-Century-Crofts, Inc., 1953), pp. 8–18. See also A. D. Ritchie, *Scientific Method* (London: Kegan Paul, 1923), Chap. II.

"inquiry," and indicate that science is a special mode of inquiry, leaving the question open for the present as to what precisely distinguishes science from other modes of inquiry.

Third, science requires *scientists.* Science does not go about the world disembodied, but is the activities of certain individuals, and the results of these activities exhibited in books, documents, and reports, on the one hand, and in inventions and objects of human ingenuity, on the other. It is common in discussions of science to regard the scientist as a kind of Absolute Knower—a completely depersonalized being who makes observations, performs experiments, describes facts, invents theories, and creates symbolic schemes, and then promptly recedes into the background and disappears. We need not point out that there is no such idealized scientist any more than there is frictionless motion or an ideal lever. Science is the work of this individual or that, living at a particular time and in a particular social group. Scientists are human beings like you and me, who, unfortunately, suffer from dyspepsia, have family troubles, and hate to pay their taxes, but who, fortunately, enjoy the company of friends, take pleasure in a good cigar, and like to go fishing. What is important is that scientists are part of science, and that the pleasures and displeasures which they experience in their scientific work are of essentially the same kind as those which they enjoy and suffer as common men in the practical relations of life and society. We shall examine the special role of the scientist as a unique personality in Chapter 8.

In the fourth place, every science is *knowledge* of some sort. By "knowledge" is meant the total collection of words, mathematical formulas, diagrams, models, and ideas, in their presumed portrayal of facts, which constitute for the scientist his understanding of the world. The scientist is seeking truth—of this there can be no doubt, though he has no positive assurance that he will ever attain it. That he even *wants* this guarantee is questionable, since if he could achieve final knowledge he would lose the challenge of unsolved problems and could look forward only to a dull existence in which there would be no mysteries to be explained. However, if he thinks not of an Absolute Truth but of the progressively developing scientific knowledge which we have today and have had in the past he can justifiably project this into the future. We know more today, and know it better, than we did a hundred years ago, and there is every reason to believe that when the next century passes we shall find the same comparison valid. This is really all that counts. We have at present a body of scientific knowledge which seems so well substantiated that it is not likely to be completely overthrown, though it may have to be trimmed at the edges and reshaped, and so challenging that its future will almost certainly consist in unlimited expansion into areas still unexplored.

Finally, every science involves *motivation.* Scientific knowledge is a value.[4] We shall have much to say about this in the future pages. Science is not merely the *sum* of a subject matter, a method of procedure, a scientist, and knowledge; for these provide merely the elements, not the cement. Science arises only because a scientist *values* knowledge to such a degree that he is willing to go out into the world and develop a method by which he can achieve this understanding. It seems hardly necessary to state that most scientists—at least most great scientists—are such because they *want* to be. *What* motivates scientists to carry on science will be our concern in later chapters. Man seeks knowledge simply because some aspect of the world or his experience is not to his liking, and he is able to envisage a state of affairs which would be more agreeable. Whether this basic motivation for all science lies in his unhappiness in living in a world which is mysterious and inexplicable, or in the pain and fear which he experiences when confronted by a world which often harms him and which he is unable to control, makes no essential difference. In the former case we define knowledge as that which satisfies curiosity and produces understanding; in the latter as that which produces a feeling of security in a hostile environment. The result is the usual distinction, common to scientific literature today, between pure or theoretical science, on the one hand, and applied science, technology, engineering, and the useful arts, on the other.

For the purposes of the present discussion this distinction is not important. Without doubt the two motives usually go hand in hand. So far as the historical origin of science is concerned we shall probably never be able to tell which was dominant in the creation of science. At this stage in our study the best procedure seems to be to place our emphasis on theoretical science, since

[4] This is not to be confused with the statement that science *studies* value, which is quite different in import.

practical science, except in its origin in primitive society, presupposes pure science. Whether we are interested only in understanding nature or also in controlling it, what we do is essentially the same. We attempt through the use of thought to construct a system of ideas, or concepts, or symbols of various kinds, which will portray in some sense the world in which we find ourselves. Inquiry is the activity of constructing such a system as this, and the resulting system, when it has the proper relation to its subject matter, is scientific knowledge or scientific truth. We can see, then, what is meant by calling science "the pursuit of truth through the method of inquiry." Because man is unhappy in the presence of the mysterious and unexplained; because he is uncomfortable when he is unable to foresee and prepare for happenings in nature; and because he is basically curious and likes to solve problems and gain understanding and insight—because of all these factors in man's make-up he sets about through the use of his imaginative and rational capacities to create a world of formulas and laws in which mystery, irrationality, ignorance, and fear are minimized. This world, in its ideal form, is a utopia in which everything can be explained and accounted for, in which man has such complete control over his environment that unforeseen events can never surprise him, and in which the forces of nature are completely at his service.

Such a picture, of course, seldom enters consciously into the mind of any scientist, and when it does it is probably vague both in its outlines and in its details. But some such ideal seems to be the guiding factor in the development of science. Certainly if we compare present-day science with the science of, say, two hundred years ago we can safely say that we are nearer such a goal today than we were at that time. We need not insist, therefore, that *every* scientist, or even that *any* scientist, have a picture of this ideal knowledge in his mind, directing his activities. We need only point out that at any given stage in the development of science, when what had been a puzzling problem becomes solved and what had been a hostile natural force yields to human control, the scientist does experience a certain kind and degree of satisfaction which causes him to say, "This is the kind of thing we are trying to do in science; this is an indication of scientific progress; what we hope for in the future is more of the same sort of thing." He then sets about to examine carefully the techniques by which he has achieved these results in order to ascertain whether they can be applied in other fields; he criticizes these techniques for their failure to produce the desired outcome more quickly; he invents and tries out still other methods and procedures in the hope of accelerating the growth of science in the direction of this unclearly defined final stage in its development. All such devices and techniques constitute the method of inquiry; inquiry is the tool by which truth is to be fully attained.

If we substitute in the following quotation the word "inquiry" for the word "knowledge," we shall have a relatively satisfactory definition of "science" at this temporary stage in our analysis:

> All these evidences together provide us at length with a general definition of science: no arbitrary phrase, but a definition which, being derived from actual common usage, is at once dependable and illuminating. Any concern or occupation sufficiently important, purposive, practical, explicit and rational, which is based on knowledge or its pragmatic equivalent, is science. This knowledge (to avoid a fussy prolixity let us call it that) may be the knowledge of natural phenomena either mechanical or vital, of emotion, will and thought, of human affairs; of ways and means, methods and procedures; of abstract relationships; of God. It is a very loose term, yet it has a sharply distinctive flavor: that of matter-of-factness and of detached rationality. This essence of the word – to return to our useful ancient conception – thus clearly implies not so much a subject-matter as an attitude of mind. And this attitude of mind, which begets a well defined habit of thought, is that which is characteristic of an easily recognizable human temperament.[5]

2. Misleading Definitions

However, the definition of "science" which we have thus far succeeded in constructing is still too broad to be useful. It does not tell us how science, as one kind of inquiry, is to be distinguished from other such modes. Many people inquire, and the techniques employed are almost as varying and numerous as are the person-

[5] Frederick Barry, *The Scientific Habit of Thought* (New York: Columbia University Press, 1927), p. 7. Reprinted by permission of the publisher. Other "general" definitions of "science" can be found in the following sources: J. B. Conant, *On Understanding Science* (New Haven: Yale University Press, 1947), pp. 36–40; J. Bronowski, *The Common Sense of Science* (Cambridge: Harvard University Press, 1958); W. S. Beck, *Modern Science and the Nature of Life* (London: Macmillan & Co., Ltd., 1957); R. B. Lindsay, *The Role of Science in Civilization* (New York: Harper & Row, Publishers, 1963), Chap. 2; W. B. Gallie, "What Makes a Subject Scientific?" *British Journal of the Philosophy of Science*, VIII, No. 3, 118–139.

alities of those who do the inquiring. The uncritical thinker inquires by using hunches, guesses, and wishful thinking; the authoritarian inquires by accepting without question the statement of one who is presumed to be possessed of higher wisdom; the mystic inquires by falling into a trance and emerging with what is claimed to be divine knowledge; even the dog may be said to be inquiring when he looks at a stranger, sniffs at his feet, and attempts to determine whether he is friendly or hostile. Our specific problem, therefore, is to determine what unique features the scientific method possesses—the features which enable us to avoid confusing it with some of these alternative methods for acquiring knowledge.

One very common attempt to define the scientific method is to identify it with what is called the "scientific spirit." Unfortunately this expression is ambiguous and may mean many things. Ordinarily by the scientific spirit we mean a certain vaguely defined attitude or way of looking at the world which is characterized by impartiality, freedom from prejudice, and respect for the criteria of truth. The scientist seems to feel to a very high degree his responsibility to the ideal of knowledge, and cautiously avoids temptations to fall into wishful thinking, or to be duped by his senses, or to accept superstition or unsupported authority as adequate knowledge. It is this strong responsibility to the ideal that restrains a scientist from publishing before he is sure of his results but impels him to do so as soon as they have been checked. This same feeling tells him that the unfettered imagination, however inviting its products may be, has no place in science; that observations obtained under conditions of haste, excitement, or fatigue, however urgently they may have been needed at the moment, are thoroughly unreliable; that "pet" theories must be verified with extreme caution because of the emotional halo which surrounds them. All these are part of the scientific spirit.[6] It can be described most adequately, perhaps, in terms of objectivity, unemotionality, rigor, and control.

Now no one, I should suppose, who knows anything about science would deny that these are essential to science. They are,

---

[6] J. Arthur Thomson calls it the "scientific mood," and characterizes it by the following elements: a passion for facts, cautiousness of statement, clearness of vision, and a sense of the inter-relatedness of things. J. Arthur Thomson, *Introduction to Science* (New York: Henry Holt & Co., 1911), Chap. I.

in fact, part of the moral code of science, and the scientist who deliberately violates them is promptly ostracized by his colleagues. But while such methods are essential they are not peculiar to science. They serve to distinguish science from pseudo science, astronomy from astrology, serious attempts to understand nature from charlatanism, and thinking which acknowledges its responsibilities to rules from thinking which does not. But much study which is under the guidance of the scientific spirit is not science.

Take the authoritarian method, for example. There is a wide range of situations in which this method is commonly used. The student accepts the ideas of his teacher; the child, perhaps unhappily, obeys the instructions of his parents; the religious devotee follows the writings in a sacred book or the teachings of a master; the art enthusiast seriously evaluates the judgment of the critic; and even the philosopher may adopt the teachings of a Plato or an Aristotle. In all of these cases the novice makes a sincere attempt to learn from the master. This method frequently demands an honest effort on his part to determine critically what the authority said, what he meant when he said it, and what reasons he had for believing as he did. Such a conception of knowing requires an initial acknowledgment that the way of authority is the way of truth, and this itself may be grounded on reasons of the heart; but once the admission has been made, the investigation into the writings and sayings of the authority may be carried on with true regard for the scientific spirit.

This is not the occasion to weigh the merits and demerits of the authoritarian method.[7] Later on we shall see that science itself uses this method, though in a restricted and controlled manner. But the authoritarian method is not the scientific method, and if we define "science" by identifying it with authority we run risk of serious confusion. So also we are liable to run into difficulties if we identify science with the scientific spirit in general; to be scientific means something more specific than to be objective, unemotional, and critical. To define "science" in these terms would be to make every serious inquiry (including every course taught in our colleges and universities) a science. We have tried to define *science*, but we have ended by defining *scholarship*.

Another very common attempt to define science identifies it

[7] For a good discussion of the authoritarian method see W. P. Montague, *Ways of Knowing* (New York: The Macmillan Company, 1958), Chap. I.

with the experimental method. The traditional picture of the scientist is that of a white-coated individual in a laboratory, surrounded by test tubes, measuring instruments, and intricate electrical and mechanical contrivances. To test scientifically presumably means to test in a laboratory. Science, according to this conception, first became scientific when it abandoned pure observation of the world of nature, and brought all objects and happenings into a building where they could be subjected to a controlled environment and their properties and behavior could be measured by various recording devices.

Since, as we have already seen, definition is to a great extent an arbitrary process, we may choose to define the scientific method in terms which identify it with the laboratory method. In proceeding in this manner we should be electing to define it in terms of the "best" or most advanced sciences of our day. But there would be some unfortunate consequences if we did this. Unless we defined the experimental method in an extremely loose and general manner, there would remain only two sciences—physics and chemistry. Astronomy could not be a science, for although it uses instruments it has no laboratory and performs no experiments. Mathematics could not be a science, for though it certainly lacks nothing so far as precision and exactness are concerned it, too, has no laboratory and performs no experiments. Only a part of biology, and that only in recent years, could qualify as a science. The same is true of psychology and of the social sciences.

Such a definition, therefore, while it might be useful for certain purposes, would restrict the field of the sciences much too narrowly. We should find as a result of our definition that there are practically no sciences. By rejecting this definition we need not, of course, deny that science uses the experimental method; we insist only that the true method of science is something more general than this and that a science uses laboratory techniques only when the subject matter and methods permit. We may even admit that a science which uses such methods makes rapid progress, hence that every science *should* strive to introduce this procedure. But we should not deny that the non-experimental and non-laboratory investigations are sciences. Having accepted the more general conception of the scientific method we can then, if we wish for further clarification, subdivide the sciences into the laboratory and non-laboratory sciences, or even into the ideal or perfect sciences and the imperfect sciences. Such a general conception of science proves to be more in conformity with accepted usage, and therefore leads to less misunderstanding in actual discourse.

3. A Definition of *Science*

Proceeding on this basis we shall now define "science" as *that mode of inquiry which attempts to arrive at knowledge of the world by the method of observation and by the method of confirmed hypotheses based on what is given in observation.* This gives us a good working definition—a definition which is relatively precise and yet in rough conformity with current usage. To be sure, in its present form it excludes both mathematics and those primarily descriptive sciences which shun hypotheses. But we shall take care of these limitations later. Furthermore, it suffers from another defect: it contains terms which require further clarification. But this is true of *all* definitions. Often we find, for example, when looking up a word in a dictionary, that the terms which are used in the definition are not clear to us and we have to look up these words also. Every dictionary attempts to explain meanings of words by relating them to meanings of other words, and the process is successful in achieving clarification only to the extent to which some words are initially clearer to us than others. In asking, therefore, that a definition use notions which are familiar to us we must recognize that this is not always possible; definition must end ultimately either in certain indefinables, or in words which can be defined by pointing to objects or by less direct methods of clarification.

Our task in the following pages will be to explain the meaning of the two terms "confirmed hypotheses" and "inquiry." If we are to understand what science is and what its role in contemporary society is we must answer two groups of questions.

First, what is the method of confirmed hypotheses? How is it motivated? From what does it start? How does it proceed? What are the rules for using it? All these are questions as to the nature of the scientific method, and we shall turn immediately to an attempt to answer them.

Second, what is inquiry? The answer to this question presupposes answers to two sub-questions. (*a*) To what extent is inquiry like the pursuits of the other basic human values, and to what extent is it fundamentally different from them? Science is a value response to the world and thus has much in common with art, religion, morality, politics, work, and play. But it can be sharply distinguished from these other modes of reacting to the world, since it deals with truth rather than with beauty, God, goodness, obedience to law, government, money, or recreational values. (*b*) How is inquiry related to these other pursuits? Human nature and society are complex combinations of science, art, religion, morality, and a host of other valuational activities which interplay to form our present civilization. This is a problem, largely, of cause-and-effect relationships. How, for example, does our science affect our religion? How does our social outlook influence our science, and how is science itself operative in determining our social values? How can conflicts between values be settled? How does science determine our general conception of what it means to live well? Question of this kind will be postponed until we have examined in greater detail the nature of the scientific method.

2

Getting The Facts

The increased worship of science in recent years has led many people to identify "the scientific" with "the factual": anything which is scientific is factual and anything which is factual is, at least potentially, scientific. Though such an identification might not survive careful scrutiny, there surely is a sense in which science begins with facts (*observed* facts) and seeks facts, in the sense of *well confirmed theories.* While there obviously are many facts which are too trivial to warrant scientific study, nevertheless these could be investigated scientifically if the occasion should arise. And there are, of course, many beliefs which are not factual; but these, it is commonly maintained, lie properly outside the field of science since they are made up largely of uncritical common-sense notions, religious tenets based on faith and affairs of the heart, and convictions of right and wrong justified only by our desire to conform to the practices of the social group of which we are members. Science, by virtue of its concern with the factual, is presumed to be superior to all this, at least so far as knowledge is concerned. The other kinds of so-called knowledge may be personally more agreeable, more comforting in times of stress, or more effective in producing desirable social behavior. But they do not meet the rigid criteria which logic has set up as the tests of genuine knowledge.

However, no important problem is really solved by insisting that science is concerned with the factual. The word "fact" is subject to a wide variety of interpretations and unless these are sharply distinguished the problem is really not clarified. Sometimes, for example, one hears people speak of "true facts" and "false facts." More refined usage forbids such expressions. A fact is not the kind of thing that can be true any more than a triangle can be virtuous or a dog can be pious or a sonnet can be equal to the square root of two. Facts simply exist or do not exist. *Knowledge* of facts, on the other hand—or at least *presumed knowledge* of facts—can be true or false. It is true when the facts claimed in the knowledge do really exist, and it is false when they do not. Thus it is the presumed knowledge which is true or false, and the facts determine whether the claim to knowledge is or is not a justifiable one.

There is, however, an important idea in the minds of those who speak of true and false facts. They are endeavoring to call attention to something which frequently escapes notice: not everything which appears to be a fact is really such. Facts are not labeled like the magic cakes in *Alice in Wonderland,* "Eat me," and the task of the scientist is not merely one of going about the world for the purpose of discovering and cataloguing all the facts which are displayed to view. His main job is to determine whether what looks like a fact is really a fact. While we cannot, therefore, admit false facts we must allow for the possibility of what might be called "pseudo facts" or "apparent facts." People once considered it to be a fact that the earth was flat, that malaria could be caused by exposure to night air, and that insane people were inhabited by devils. We now consider these to be pseudo facts, and we have replaced them by genuine facts. In this opinion we may, of course, be mistaken, and our grandchildren may laugh at our willingness to be duped by facts which were only apparent. The ease with which we are often deceived means, simply, that the process of determining what is a fact and what merely appears to be a fact is complicated and involved. It is, indeed, science itself.

1. Science and Facts

In what sense, then, is science concerned with facts? There are at least three.[1] In the first place, facts are the starting point for all scientific investigation. "Any scientific idea arising in the mind of a scholar is based on a concrete experience, a discovery, an observation, or a fact of any kind, whether it is a physical or an astronomical measurement, a chemical or a biological observation, a discovery among the archives or the excavation of some valuable relic of an earlier civilization."[2] Science never arises unless there is something which, because it is real and inescapable, must be fitted into the general pattern of things. Our response to this new datum may be automatic and unconscious, in which case no scientific reaction arises because there is no problem. An experience becomes a problem only when it generates a conflict either between behavior responses or between opposing interpretations.[3] For pure science problems are occasions when something happens which cannot be readily explained. That which requires explanation is not necessarily the unusual; science is fundamentally much more concerned with the regularities of nature, and becomes interested in the breaks in this uniformity only when they occur as surprises, that is, as events which had not been foreseen. But whatever may be the concern of science, unless something happens there can be no science. That which happens is the fact with which science begins. For primitive man, as for us today, the sun rose in the east and set in the west, people who had been exposed to night air frequently developed malaria, and insane people sometimes talked to themselves and acted in an unconventional manner; these were facts for him as they are for us. But our

---

[1] Cf., Jevons' statement that every science contains facts of four different kinds. W. S. Jevons, *Principles of Science* (London: Macmillan & Co., Ltd., 1907), p. 525.

[2] *The Philosophy of Physics* by Max Planck. Copyright, 1936, by W. W. Norton & Company, Inc., New York, N. Y. Copyright Renewed, 1964, by Mrs. W. H. Johnston. Reprinted by permission of the publisher. P. 89.

[3] Whether science "begins" with *facts* or with *problems* has been debated at great length. Formulated in this way the question is not capable of being answered for science as a whole, since problems differ. Sometimes a scientist notes a fact which he might not have observed had he not been thinking about a certain problem; on such occasions problems may be said to precede facts. On the other hand, a fact may force itself on the attention of a scientist simply because of its great intensity or its unusual character; in this type of case facts not only precede problems but actually create them. For a good discussion of the relation between facts and problems see John Dewey, *How We Think* (2d ed.; New York: D. C. Heath & Company, 1933), Chaps. VI, VII.

interpretations of these facts are different; we start from the same facts but we arrive at different explanations. The problem of science is essentially one of accounting for something which in a very clear-cut sense does exist. We never ask the question *Why?* unless there is something mysterious for which we are trying to account; we never look for a cause unless we have something which we presume to be an effect. Science must always begin with a statement like "I am now observing such-and-such," or "Here is something to be accounted for." To deny that science begins with facts is to destroy the very reason for its existence.

But there is a second sense in which science is concerned with facts. Every theory or hypothesis is tested *by means of facts.* These facts are of essentially the same kind as those with which science begins, but they are discovered only after a theory has been devised, and they are used for the purpose of checking or proving the theory. Consider, for example, a simple mystery story. Suppose in the attempt to determine who committed a certain murder the investigator has reached a point where suspicion centers on a Mr. Wilkinson. It is readily seen that Wilkinson could not have been responsible unless he had been at the scene of the crime at the time when the act was committed; the theory of Wilkinson's guilt requires this as a necessary consequence. If it can then be shown that Wilkinson was actually present at the required time, this fact becomes evidence in support of the theory that he was the guilty man. Facts of this kind play a very important role in science. They are different from the preliminary facts in the sense that we know before we have discovered them that they ought to be there—at least they ought to be if the theory which has suggested them is correct. Often such facts have to be hunted for in out-of-the-way places, and frequently they have to be produced through experimental techniques. If we find them, our theory is confirmed; if we fail to find them, our theory remains in the balance or must be replaced by another.

We have seen thus far that there are two senses in which facts enter into science—facts which are given as the starting point for investigation, and facts which are predicted through the use of theory and later observed directly or produced experimentally. But there is still a third meaning of the word "fact" as it is used in science. A theory which has been adequately tested is sometimes called a "fact." This is clearly indicated by the way in which the words "hypothesis," "theory," and "fact" are commonly employed. They are arranged in a series in such a way as to indicate that while a hypothesis is a mere guess, a theory is a guess which has been confirmed up to a certain point, and a fact is a theory which has been so completely confirmed as to remove it substantially from the realm of conjecture. Usage here is not entirely consistent, and could be improved. Since both hypothesis and theory represent degrees of knowledge, it would seem preferable to replace the word "fact" in this trio by the expression "confirmed belief." Then we could speak of the development of knowledge as passing from an initial stage of hypothesis or mere conjecture, through a second stage of theory or credibility, to a final stage of truth or practical certainty. The use of "fact" in this connection is based on a confusion between true belief which refers to fact, and the fact which is referred to by the belief. If our knowledge about molecules is highly confirmed there are molecules; but the molecules themselves should not be confused with our knowledge about them even though our conviction of their existence is presumed to be well established.

As a result of the above distinction we are now in a better position to see what the job of science really is. Its task is to discover facts. In some cases this is a simple matter; many facts impose themselves upon us merely because we cannot avoid them. We live in a world of facts and our life is simply a matter of adjusting ourselves to these conditions of living. But the scientist is instilled with a fervent desire to discover more and more facts. If he wished to proceed very inefficiently he could simply go about in nature, looking here, there, and everywhere, and waiting, like Mr. Micawber, for something to turn up. Ultimately, if he looked far enough and waited long enough, he would presumably discover all the facts there are. But nature has provided him with an exploratory tool which enables him to proceed much more efficiently. By virtue of the activity of his mind he is able, on the basis of the facts which he has already discovered, to make controlled guesses as to what facts he is likely to find if he looks further. Frequently these guesses involve the anticipation of what he will observe at some later date. If he waits he can, of course, test many of these guesses by later observations; and if he meets with success they cease to be mere theories and become factual statements. But others of his guesses cannot be tested in such a direct way. This limitation may be because the facts in question are too small to be observed, or too remote in time or space, or not endowed with the usual perceptual qualities. Electrons, for example, are incapable of being perceived, and we must accept them on the basis of indirect evidence. We can never know positively that such things exist, but we can, by combining the information obtained from earlier observations with that obtained from later ones, and by examining the mutual consistency of what we have observed and the guesses which we make, increase the probability of our knowledge to the point where it amounts to practical certainty.

As a result we can see science as a monumental task in which we endeavor to increase the area of discovered fact by devising, through the activity of our minds, a system of symbols or ideas which actually goes beyond the realm of observed fact, but not, we hope, beyond the realm of facts which we shall discover in the future, or beyond the realm of facts which we could discover if we were not so restricted in the use of our senses. Since our system of ideas can never be compared, element for element, with our system of facts we can never be sure that our conjectured knowledge is correct. It can, at best, be merely probable and is subject to continual modification and revision as we progressively explore more and more of the world of fact, and as we increase the sensitivity of our observational instruments. Some discrepancy between knowledge and fact seems inevitable. For example, our knowledge is, on the one hand, always *less than* the facts, since at any one time there are many facts which have not yet been discovered; but our knowledge is, on the other hand, always *ahead of* discovered facts, because we have anticipated by means of guesses what nature is later going to reveal to us. The final goal of science is to make knowledge "fit" the facts in all its parts and in all its details.

Our problem at this stage is to determine how the scientist obtains the facts in terms of which the validity of his knowledge is to be measured. Unless he has facts to start with he cannot even begin his enterprise, and unless he has more facts by which he verifies his excursions into thought he has no way of determining whether these theories are true anticipations of what nature is going to reveal or mere figments of his very active imagination. We may speak of this initial state of science as involving two processes: *getting* the facts, and *manipulating* the facts. A few brief remarks concerning each of these two processes will now be in order.

2. OBSERVATION

The basic method for acquiring facts, both in science and in the life of common sense, is observation.[4] The world, at least in the early stages of our development, is simply what we see, hear, smell, taste, and touch. That it does not remain merely this for most of us is due to the creative and exploratory activity of thought. But this theorizing process is not our concern at the moment. All such activity presumably arises only because of our interest either in extending the area of sense experience or in explaining such experience. Consequently the observed facts are primary both in the sense that they are earlier in time, and in the sense that all verification of hypotheses must be in terms of them. That we do not need all five senses in order to survive is clearly demonstrated by those individuals who have had the misfortune to be deprived or one or more. Yet it seems clear that such individuals have a more restricted experience than do those of us who are in possession of our full senses, and it seems safe to say that anyone who was deprived of all of them could not long survive.[5]

To one who has never thought about the matter the process of observation seems exceedingly simple. Seeing involves nothing more than opening our eyes and receiving the stimulus sent out by the visual object; to hear we need only listen; to touch we have only to be in contact with the object whose nature we are

[4] For popular references on *observation* see M. D. Vernon, *The Psychology of Perception* (Baltimore: Penguin Books, Inc., 1962); W. I. B. Beveridge, *The Art of Scientific Investigation* (New York: Random House, Inc., 1957), Chap. VIII; and H. A. Larrabee, *Reliable Knowledge* (Boston: Houghton Mifflin Company, 1945), Chap. IV. For more advanced and analytic studies, see H. H. Price, *Perception* (London: Methuen & Co., Ltd., 1961), and R. J. Hirst, *The Problem of Perception* (New York: The Macmillan Company, 1959).

[5] Eddington points out that in a highly developed science, such as physics, where machines do much of the observing for us, the actual scientific observer can get along with a very restricted range of sense organs. He "has one eye (his only sense organ) which is color-blind. He can distinguish only two shades of light and darkness so that the world to him is like a picture in black and white. The sensitive part of his retina is so limited that he can see in only one direction at a time . . . we have left the observer power to recognize that a pointer coincides with a gradation on a scale." A. S. Eddington, *New Pathways in Science* (Cambridge: Cambridge University Press, 1935), p. 13. Reprinted by permission of the publisher.

trying to discover; and so on for the rest of the senses. The possibility of error in observation seems very remote. Seeing is believing, and justifiably so. Argument ceases if we can prove that we "heard it with our own ears." If food tastes sweet to us no vote of authorities to the contrary can convince us that we are wrong. Observation is the final court of appeal. What we observe seems to be fact, and never pseudo fact or apparent fact.

Experience, however, soon convinces us to the contrary, and a very superficial analysis of the nature of the observational acts explains to us how it is not only possible but, in a sense, inevitable that error should occur with great frequency. Let us take vision as an example; what happens in the case of the other senses is similar.

In the first place, the stimulus—in this case the light rays—must pass from the object to the eye of the observer. This always involves some spatial separation, since we cannot observe an object which is in contact with our eyes. But because of this separation distortion is bound to occur. The most common illustration is the apparent merging of the railroad tracks as we stand between them and look into the distance. But perspective is also important in the sense of position; for example, the apparent shape of the surface of a table changes as we walk around it. There are other disturbing influences, such as the nature of the light which illuminates the object, and the nature of the intervening medium; for instance, whether the air is clear or filled with smoke or fog, or whether there is between the observer and the object a source of heat which gives rigid objects wave-like boundaries, as in the observation of the edge of a building on a hot day. Observation through media other than air produces even more extreme modifications; for example, a straight stick appears bent when partially immersed in a glass of water.

In the second place, we must take account of the modifications due to the organism of the observer. After the stimulus is received at the surface of the eye it must be transmitted through the eye, along the optic nerve, and into the brain. The transformations which are due to color blindness, astigmatism, double vision, and other abnormalities play an important role here. I myself, for example, am very uneasy when having my eyes examined; knowing that the optician who is attempting to correct my vision must use his own eyes to make the test, I continually ask myself what assurance I have that he is himself seeing correctly. But because vision is a physiological function it is subject to all the distortions which are due to the condition of the organism; vision cannot be relied upon when we are fatigued, or ill, or under the influence of drugs of certain kinds. Many of the things which we "see" in our dreams are due to the disturbances which originate within the organism.

But even this is not all. Before the stimulus can enter consciousness it undergoes all sorts of transformation due to the "set" of our minds. Frequently we see what we expect to see, or what we fear we shall see, even though the object itself is not present. What we see in any given case contains elaborate additions drawn unconsciously from our past experience; while there may be something that really comes in from the outside, the interpretation which is projected "from the inside" often determines to a great extent what we are seeing. The velvet which "looks" soft, and the winter scene which "looks" cold are clear-cut examples of this spontaneous addition of remembered conceptual material to the objects which we are at the moment observing. On a certain occasion I was in need of a pencil and, finding none in my pocket, borrowed one of the mechanical type from a friend. Much to my amazement, the lead bent under the pressure when I started to write with it. The pencil was a "trick" one carried by my friend in order to fool gullible people like me who believed that what looked like a pencil must be a pencil; it was filled with semi-hard rubber instead of lead. The unconscious addition of conceptual material drawn from past experience is illustrated in the classic examples given in Figures 1 and 2. In Figure 1 the object may appear to the observer as a tunnel into which he is looking, as a framed picture, or as the frustrum of a pyramid which he is viewing from the top. The staircase in Figure 2 may be seen either above or below the line of horizontal vision when the one or the other is suggested to the observer. Excitement and other emotional disturbances influence both what we see and the way in which what we see is interpreted. The law courts provide many examples of widely varying descriptions of a certain happening as seen through the eyes of different observers, all of whom were actually present and in a position to see what occurred, but are unable, because of emotional overtones created by the occasion, or because of previous biases, to agree as to what really took place.

Such psychological factors, too numerous to mention even in outline, are some of the most important sources of error in observation.

In view of these considerations the problem of getting the facts through observation becomes much more complicated. The wonder is that we ever get what might be called reliable knowledge. Fortunately most of the errors can be eliminated by proper attention to the conditions of observation. The scientist achieves his results simply because he has become a trained observer. He knows what the disturbing factors are and he guards against them

FIGURE 1 FIGURE 2

PERCEIVED OBJECTS INTERPRETED BY CONCEPTS

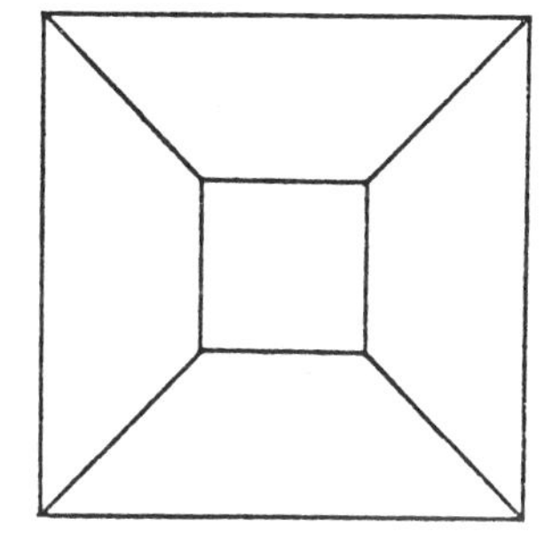

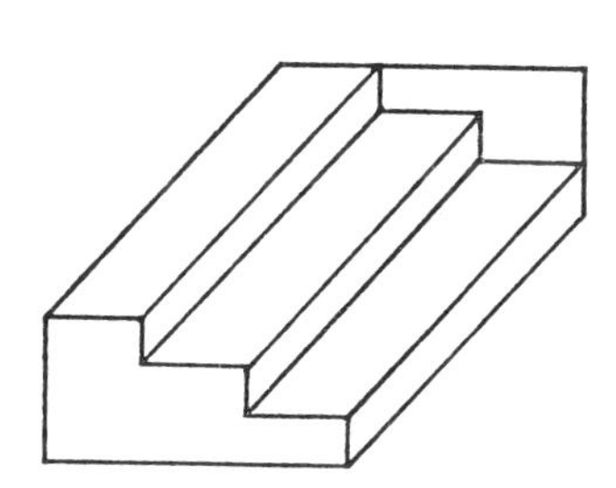

by setting up ideal conditions for observation. He ascertains that his object is properly lighted and that there is no distorting medium between himself and the object, that his eyes are normal, that he is not himself ill or fatigued at the time of observation, and that he does not unconsciously add to the observed object or modify it to conform with what he hopes to find. Under such ideal conditions the likelihood of error is very much reduced, and the scientist can rest assured that he is probably getting the facts. There always remains, however, the possibility of some correction due to the acquisition of later information. When the scientist begins with observations he does not take these as the final court of appeal; he simply accepts them at the moment and sets the theorizing activities into operation upon these as a basis.

This would be the logical point in our discussion to examine the question of *introspection* and its role in the psychological and social sciences. But space does not permit this departure. This is a rather serious omission since, although there is heated debate among behavioral scientists, particularly psychologists, as to the validity of the method, all of them, whether they do or do not accept its validity, continue to use it in the pursuit of their studies. The problem about introspection for psychologists is how to reconcile its use with the absence of any theory which will justify it.[6]

The claim is often made that the scientist draws his facts not only from observation but from reports of other scientists. While this is most certainly true, it can hardly be called a new and different source of information. For reports from other observers are merely indirect observations on the part of the scientist himself. A report is a mediated source of information, and plays much the same role in science as do the data obtained from recording instruments of various kinds. No scientist can observe all the facts which he needs in his investigations; he cannot be in all the required places and he does not live long enough. Consequently he must supplement his knowledge by reading books and listening to other individuals who have observed the facts in question. In this sense the scientist uses the authoritarian method, discussed on p. 18. But the scientist is also limited in his source of data because of the many ways in which his sense organs are restricted in range and accuracy. He therefore employs telescopes, microscopes, spectographs, ammeters, balances, thermometers, and numerous other recording devices to extend the scope and precision of his senses. Now the exact role and function of reports in science is not always clearly understood. In particular, the close similarity between the reports of human observers and the reports of mechanical recorders is often overlooked. We should note that observation is involved in both cases. When the scientist reads or listens to the words of his colleague he is seeing ink on the printed page or hearing noises emitted through the mouth of his associate; in the same way, when he employs an instrument, say, a telescope, he observes an image as "directly before" him. Further-

[6] A few suggested references are *Theories in Contemporary Psychology*, ed. by Melvin H. Marx (New York: The Macmillan Company, 1963), pp. 235–236; C. C. Pratt, *The Logic of Modern Psychology* (New York: The Macmillan Company, 1939), pp. 48–57; and P. W. Bridgman, *The Way Things Are* (Cambridge: Harvard University Press, 1959), Chap. VI. An excellent reference is Peter McKellar, "The Method of Introspection," *Theories of Mind*, ed. Jordan M. Scher (New York: The Free Press of Glencoe, 1962), pp. 619–644.

more, what the scientist is primarily concerned with in each case is not that which he directly observes; his main interest is in what he can legitimately infer on the basis of what he sees and hears and what he knows about the nature of the reporting instrument. If he is correctly informed about the observational habits of his associate and his reputation for communicating information reliably he can infer that the objects as seen by his colleague are such as he, himself, would see if he were subjected to the same conditions of observation. Similarly if he knows enough about the telescope and the way in which magnification takes place, he can infer from the image seen in the aperture something of what he would himself see if he were very much nearer the object. Thus the scientist observes something directly for the purpose of learning through inference about something else which he cannot observe directly. This inferential knowledge is frequently a source of error. He sometimes accepts the word of unreliable witnesses, and he frequently takes over the results of imperfectly functioning recording instruments. In any case the instruments are utterly valueless unless he knows how they function normally and the kinds of errors to which they are subject.

3. Experimentation

Let us now turn to the second of the preliminary processes in science—the manipulation of the facts.[7] The principle here is simply that by the transformation of facts we can produce new facts. Science has not only *observ*atories but *labor*atories. The activities involved are those which are commonly called "experimental." In the best sense of the word, experimentation occurs rather in the testing stage than in the initial stage of discovery, where we now find ourselves. But in the broad sense any operations which involve inserting one's finger into nature for the purpose of chang-

[7] Discussions of *experimentation* range from very popular collections of actual scientific experiments, through attempts to distinguish experimental from non-experimental sciences, and end with complicated theoretical designs for experiments. Following are suggested readings drawn from this extensive area: Ritchie Calder, *Science in Our Lives* (New York: Signet Key Books; New American Library of World Literature, Inc., 1954), Part Three; Beveridge, Chap. 2; Claude Bernard, *Introduction to the Study of Experimental Medicine* (New York: Dover Publications, 1957), Part One; Jevons, Chap. XIX; and Ernest Nagel, *The Structure of Science* (New York: Harcourt, Brace & World, Inc., 1961), pp. 79–90.

ing things are experimental. These manipulatory acts have their foundation in the random grasping, pinching, tasting, and throwing activities exhibited by a child when he is given a new toy. If we could read the infant's mind we might discover that he is trying to find out how the object will behave under varying circumstances. This is essentially what the scientist does in his effort to increase his knowledge of an object. Sometimes the scientist has a definite idea in advance as to how the object will act under novel conditions; he then performs what is commonly called an "experiment for testing." On another occasion he may have no such advance knowledge and merely introduce a change into nature to see what will happen; this manipulation act is usually called an "experiment for discovery." The activity which is involved is much like that of children when they play the game of "supposing." Suppose you had a million dollars, what would you do? But while children do this merely for amusement, scientists engage in such activities in order to learn more about nature; if we knew what a poor man would do with a million dollars we should have real insight into his character. Were we actually to give such a man the money and then observe what he did, we should be employing the experimental method in the strict sense of the word.

The attempt has been made by Claude Bernard,[8] following Bacon, to distinguish experiments for discovery from experiments for confirmation by saying that in the former we are merely "listening to nature," while in the latter we are "asking nature questions." The analogy is suggestive since in listening we are usually passive whereas in asking questions we are certainly active. But I hope that the brief analysis of observation given above has dispelled the illusion that any careful perceiving of objects in the world is in any sense passive; we must be continually on the alert for misinterpretations, malfunctioning of the sense organs, and abnormal environmental media. However, the distinction between the two modes of experimentation—which is an important one—can be maintained on other grounds. In both cases we are active in the sense of "putting our finger into nature" and bringing about a state of affairs which would not normally occur without our intervention. We thus prod nature and man in order to in-

[8] Bernard, p. 22.

duce them to disclose some of their dispositional properties. But the distinction between the two forms of experimentation can now be explained in terms of the *kinds* of questions asked in each case.

In experiments for discovery the questions are vague, and in no sense are they *leading* questions. In fact there may be no questions at all and the experiments may fall into the category of random manipulations performed in the hope that by pure chance something significant may turn up. I was told by a soap manufacturer that it was not uncommon some twenty-five or thirty years ago for men of his trade to employ in their research laboratories young and inexperienced chemists, whose job was to try, more or less at random, one mixture after another, with the unlikely expectation that a new formula for soap might emerge. Thomas Edison's invention of the electric light was the result of a large number of more or less unguided trials; by contrast the work of Irving Langmuir in the same field proceeded under the direction of a hypothesis. Many drugs are discovered through exploratory experiments; a compound having certain desirable curative properties is known also to have unpleasant side effects, and the attempt is then made, sometimes in a random way, to eliminate the unwanted properties without losing the valuable ones. W. E. Berg reports that the development of photographic emulsions has been the result not of hypotheses but of practical trial. "In spite of all . . . efforts, the preparation of the light-sensitive medium, the photographic emulsion, is essentially an empirical matter and improvements have come as a result not of predictions by the 'backroom boys' but of the bottle-from-the-shelf approach."[9]

In the natural sciences in general we often ask nature such unpointed questions as what will happen if we heat you, freeze you, pass a current through you, place you in sulphuric acid, expose you to light, fertilize you, breed you in selective ways, irrigate you, and so on. In the behavioral sciences we are even today asking the public what will happen if we lower income taxes without balancing the budget, what will be the result of passing a law which forbids racial discrimination in places of public accommodation, and even (in view of the uncertain outcome of psy-

[9] W. F. Berg, "Latent-Image Formation and Chemical Sensitization," *Scientific Monthly*, Vol. 80, No. 3, 163.

chiatric treatments) what will happen to a given individual if we psychoanalyze him.

But in experiments for confirmation, since we already have tentative hypotheses in mind, we can ask questions which are pointed and actually contain a suggested answer. We can "ask" water whether it will boil when it is heated sufficiently, whether it will expand when it is frozen, whether it will conduct electricity if an attempt is made to pass a current through it, whether it will dissolve salt when this is placed in it and the mixture is stirred, and so on. In the biological sciences, e.g., botany, we can "ask" a certain plant whether it will not die if deprived of water, whiten if deprived of light, become more luxuriant if fertilized, wilt if exposed to excessive heat, and so on. In the behavioral sciences, as history reports, we once asked Americans what they would do if they were forbidden to buy intoxicating liquors (in the Volstead Act), what they would do if offered the "opportunity" to avoid starvation by accepting degrading, low-paid jobs on WPA projects (as happened in the nineteen-thirties), or what they would do if Germany were to invade Poland (as happened in 1939). In all cases we got answers—sometimes those which were expected, sometimes not. The prohibition amendment did not provide a solution and had to be repealed; WPA helped to alleviate the situation which created it; the first World War solved its problem temporarily, but not permanently. (It should be noted that when we speak of an experiment as working or failing in the behavioral sciences we are moving into the area of applied science. For we were trying out these cases in order to cure social and political ills by means of instruments whose effectiveness had not previously been ascertained; we were not attempting to determine how people behave when deprived of intoxicating liquor, or when starving, or when faced with an international atrocity. But we did, as a matter of fact, learn much about man's behavior through these experiments, and they greatly enlarged the realm of data relevant to further sociological theorizing. Bacon suggested that experiments carried on for the purpose of gaining new insights be called "experiments of Light," and those performed in the hope of improving man's condition on the earth "experiments of Fruit.")[10]

[10] Francis Bacon, *The Philosophical Works of Francis Bacon, Novum Organum* (London: George Routledge & Sons, Ltd., 1905), Aphorism xcix.

The question as to whether there can *be* any social experiments, in the strict sense of the word, is an issue which is too complicated to be considered here.[11] In the broad sense of "experimentation," all the social processes of passing laws, changing political administrations, raising interest rates, abandoning towns, moving from the country into the city, joining the common market, divorcing, psychoanalyzing, marrying, employing men, and beautifying cities are just as much experimental as condensing, magnetizing, stretching, chemically combining, pulverizing, fertilizing, dissecting, grafting, changing environment, and other physical changes. The fact that social changes cannot be carried on under the same controlled conditions that natural changes can, does mean that in the area of the behavioral sciences experiments are less productive of precise and highly confirmed information than in the realm of the natural sciences; but it does not mean that experiments are not being performed and cannot be performed. It is therefore quite legitimate to speak of the prohibition amendment and the WPA as social experiments, though the information we obtained from them was much less exact and complete than we wish it could have been.

The properties of objects which are disclosed through experimentation are commonly called "dispositional properties,"[12] in contrast to actual or real properties. Both may be called "properties" since they may be attributed to objects in much the same way. But the latter are properties which we discover in objects by a more or less simple inspection, while the former can be uncovered only by changing the circumstances in which we find the objects and noting how they behave after these changes have been made. Dispositions are potentialities, capacities, tendencies, inclinations, powers, abilities. While there are important differences in the ways in which these several terms are used they have

---

[11] Nagel, pp. 450–459.

[12] Though Gilbert Ryle did not invent this term he has given one of the clearest statements of what he calls the "logic of dispositional concepts in general": "When we describe glass as brittle, or sugar as soluble, we are using dispositional concepts, the logical force of which is this. The brittleness of glass does not consist in the fact that it is at a given moment actually being shivered. It may be brittle without ever being shivered. To say that it is brittle is to say that if it ever is, or ever had been, struck or strained, it would fly, or have flown, into fragments. To say that sugar is soluble is to say that it would dissolve, or would have dissolved, if immersed in water." Gilbert Ryle, *The Concept of Mind* (New York: Barnes & Noble, Inc., 1949), p. 43. Reprinted by permission of the publisher.

in common the feature of something which *will be* exhibited in the future if we *do* certain things, or *would be* exhibited if we *should do* certain things (even though we may not actually do them). Dispositional properties are frequently indicated by words ending in "-ible" or "-able"—*soluble, combustible, malleable, fissionable, edible, adaptable, reasonable, sociable,* and so on. No doubt it would be difficult to draw a sharp line between real properties and dispositional properties. We should be tempted to say that the whiteness of salt is a real property while its solubility in water is dispositional, because in order to detect this property we should have to place some salt in water and stir it or heat it. But in order to detect the whiteness of salt we have to turn the light on it, for it is not white in the dark. And in order to observe any object we must focus our attention on what is presented to us. Chemists often make a distinction between the physical and the chemical properties of a substance. Presumably the basis for this differentiation lies in the fact that the physical properties—taste, odor, color, state (liquid, solid, gas), solubility, and density—are determined by observation while the chemical properties—reaction to light, to shock, and to other chemical substances—can be determined only by experiment.[13] That this distinction cannot be maintained is illustrated by the fact that solubility and density are not strictly observational properties.

In spite of the difficulty we experience in making a sharp distinction between dispositional and real properties, the contrast between the two is useful for science. The distinction is important in the social sciences, and especially in psychology. Psychologists of the behavioristic school are disinclined to talk about the mind or the self because their properties are not open to observation but must be determined through introspection. This substitute method implies, however, that the usual observational checks through corroboration by other people are not available. Each person's mind is known only to himself, and there is no method by which his introspective methods can be validated. Many psychologists, therefore, prefer to define the subject matter of psychology as behavior rather than as mind or ego. Behavior is open to public inspection and we can agree on how people behave just as we can agree on how minerals and plants and animals

---

[13] As stated by John R. Lewis, *First-Year College Chemistry* (7th ed.; New York: Barnes & Noble, Inc., 1953), pp. 3–4.

behave. Psychologists of this school tend, consequently, to talk about dispositions to behave rather than about images, ideas, decisions, feelings, dreams, and hallucinations. Gilbert Ryle claimed to define all subjective concepts in terms of behavior and dispositions to behave. Similarly in sociology, economics, and political science dispositional properties play an important role. One need only draw up a list of "ble" terms to realize this: *unemployable, sociable, marriageable, dependable, saleable, transportable, financially insoluble, negotiable, eligible* (to vote), *inviolable* (rights), *debatable*, and *undefeatable*.

Before leaving the subject of experimentation we must raise an important issue. Experiments, we have agreed, are alterations introduced into nature and society for the purpose of revealing "dispositions" on the part of objects and people to behave in ways other than those disclosed to direct observation. Sometimes these changes involve merely the use of instruments, introduced into the physical or social medium in which the subject matter of our study is located. They are not commonly called "experimental" since they ordinarily do not destroy or seriously transform the objects or people of whose behavior the recordings were made; we may say, therefore, that the use of instruments or the mere presence of the observer does not "transform" its objects but merely multiplies the "appearances" which the objects exhibit under different circumstances.

But does not the presence of a recording instrument, or even the presence of the observer without any instrument, often "transform" the object in ways which may be unknown to us at the time or in ways which, even if known to us, are more or less completely disregarded? Let us first take cases in which the presence of an instrument is known to transform its object. If we measure the temperature of a liquid in a flask by placing a thermometer in it, we know that unless the temperature of the thermometer itself, prior to its immersion, is exactly the same as that of the liquid the reading will be inaccurate; if the thermometer is warmer it will raise the temperature of the liquid and if it is cooler it will lower the temperature of the liquid. Thus we do not get the "true" temperature of the liquid but its temperature as modified by the presence of the thermometer. The classic example of an operation heretofore supposed to be nontransforming is the observation of a photon. Heisenberg showed that the position and velocity of a photon cannot both be determined at the same time with a high degree of accuracy. Since the particle is very small a very short wave must be used to determine its position. But a radiation of short wave has a high momentum, since wave length and momentum are in inverse ratio to each other. Consequently, the electron struck by this wave will be pushed from its position, and its motion altered. If this difficulty is avoided by using radiation of long wave, the momentum is decreased and there is less recoil, but now it is no longer possible to tell precisely *where* the electron is. Hence because our only recording instruments prevent us from obtaining the complete information of the state of the electron—information which is necessary to predict its future path—a transforming act is always present, and we can never "really" know our object in this respect.

The problem of unconscious transforming operations in observational and experimental situations runs the entire gamut of the sciences. In the case of the physical sciences the transforming operations occur in the instrument rather than in the observer himself. Exceptions can be found in cases where the heat of the observer's body may affect the result, motions of his hands may disturb the stability of the system with which he is dealing, and chemicals from his hands (e.g., perspiration) may change the nature of an object if he merely picks it up and moves it. Where transforming operations do occur in the instrument, allowance can usually be made for this fact and the disturbance reduced to a minimum. If we observe snowflakes falling on a piece of dark wool cloth, then on a sheet of ice, and finally on a hot griddle, we can easily explain and allow for the different properties which are exhibited by the observed objects. Where the transforming operations are very significant, the instrument which produces them is abandoned; for example, one would not normally measure the size of an ice cube by using a red-hot measuring rod.[14]

In the biological sciences, especially the botanical sciences, the situation is much the same as it is in the physical sciences. Of course the fact that the subject matter is living organisms greatly reduces the number of possible experimental acts that may be performed; most of the operations of physics, applied to organisms, would kill them immediately and the value of the experiment would be greatly reduced, if not completely nullified. But in the

---

[14] H. Levy, *The Universe of Science* (New York: Century Co., 1933), p. 66.

case of experiments on animals, especially those which we consider to be in some sense conscious of our presence when observing them, disturbances of a significant sort often occur without our being aware of them. One of my colleagues in ornithology protests because many zoologists study wild bird life by catching wild birds, putting them into cages and then observing and describing their behavior. But putting a wild bird into a cage may significantly change its behavioral pattern; the only alternative is to go into the country and engage in "bird watching"—an activity which is often looked upon by ornithologists with scorn because it is either not scientific at all or merely a highly diluted form of fact-gathering.

It is in the area of the behavioral sciences where the presence of the observer most often transforms the character of the observed objects. Here instruments are seldom used, unless one wishes to call questionnaires, voting machines, interviews, and psychological tests instruments of a certain kind. I mention only a few random examples of situations in which the presence of either the observer or the "instrument" may well bring about a significant transformation in the character of the object observed. An anthropologist goes to an isolated but inhabited island where there are seldom visitors; he then lives with the people over a period of time, notes their customs and manners, and writes an account of the culture of this tribe. But how does he know that his very presence in their group did not lead them to modify their normal behavior and thus make the results of his study inaccurate? A student taking an examination may often be unable to record correctly what he knows simply because the tension which is produced by his knowledge of the effect of the outcome on his future life may prevent him from recalling accurately and writing clearly. The value of the famous Kinsey reports[15] is considerably lessened by virtue of the fact that the only way to collect data is to find individuals who are willing (and sometimes even anxious to the point of fabricating totally unreal situations) to write and talk about their own sexual behavior. We all know individuals who consider their sexual life nobody's business but their own, and they would obviously refuse to take part in such

[15] A. C. Kinsey *et al.*, *Sexual Behavior in the Human Male* (Philadelphia: W. B. Saunders Co., 1948); and A. C. Kinsey *et al.*, *Sexual Behavior in the Human Female* (Philadelphia: W. B. Saunders Co., 1953).

a survey; on the other hand, we are acquainted with people who continually talk about their sex life, even to the point of boasting about it, and these individuals would welcome an opportunity to take part in a scientific study in which they provide data. Psychiatrists know that interviewing a patient transforms the patient; in fact this is often the purpose of the interviewing activity. But when they are engaged in psychoanalysis they usually sit at the head of the couch on which the patient is lying in the hope that since they cannot be seen their presence will not significantly determine what the patient reports. In the case of street interviews or house-to-house canvasses, the attitude of the interviewer and even his language must be carefully watched in order to prevent falsification of the data. To call an intellectual person a "highbrow," a believer in democracy a "socialist," and a Negro a "nigger" is often to introduce a transforming factor of the utmost significance.

Another method in the behavioral sciences, commonly urged by historians (if they may for the moment be considered scientists), has been called variously, "empathy," "*Verstehen*," "sympathetic understanding," and "interpretative explanation."[16] It involves an act on the part of the scientist which might be called that of "taking the role of another."[17] Historians often argue that we can better understand situations in the past if we can project ourselves back into the places of those who lived in these times, thought in the manner of their compatriots, and enjoyed and suffered the experiences which fell to their lot. For example, how explain Napoleon's behavior after his defeat at the Battle of Waterloo? The best way to do this is to put yourself in his place and try to imagine how he thought and felt on this occasion, and how you, granting these same thoughts and feelings, would have behaved. This, of course, offers no kind of objective test, but it does, according to these historians, contribute that element of "understanding" which seems to be lacking in a completely objective confirmation.

---

[16] Theodore Abel, "The Operation called *Verstehen*," *Readings in the Philosophy of Science*, ed. Feigl and Brodbeck, pp. 667–687; Nagel, pp. 480–485; and *Philosophy of the Social Sciences*, ed. by M. Natanson (New York: Random House, Inc., 1963), *passim*.

[17] G. H. Mead, *Mind, Self, and Society* (Chicago: University of Chicago Press, 1943).

3

The Language of Science

It should be clear from our discussion up to this point that we have gone but a short way in our attempt to understand the nature of science. We have shown how to gather facts. But to have facts is not science. We could suppose an individual going about in the world and observing a wide range of interesting phenomena; we could even imagine him setting up a laboratory and spending his days playing with various gadgets designed to produce objects which had never before existed, and events which had never before occurred. But there would be no science unless he made the attempt to record his observations, and to systematize and classify them. In fact, in view of what we have seen about the unconscious intrusion of conceptual material drawn from his past experience into the very act of observation itself, we should have to admit that he could not observe anything without using a language of some sort. If he remembered any of his previous observations he must have symbolized them at least to the extent of forming memory images of them, and it is hardly conceivable that he could avoid the more complex activities of the mind involving conceptualization and the formation of general notions. Mere awareness of the world is not science; indeed, it is not knowledge at all. The world enters properly into knowledge only when it is described, interpreted, and explained, and these supplementary activities require the use of symbols. Whitehead suggests that there is a great difference between direct knowledge and symbolism. "Direct experience is infallible. What you have experienced, you have experienced. But symbolism is very fallible, in the sense that it may induce actions, feelings, emotions, and beliefs about things which are mere notions without that exemplification in the world which symbolism leads us to presuppose."[1] The function of acquaintance is simply the production of that givenness which is assumed in all knowledge. We have examined the processes by which the world of science is given. If we were to call the activities which we have examined "science," we should deprive that word of three important aspects of its meaning.

1) "Knowledge" of this kind would not have the public and communicable property which we think of as belonging to science. Even if individual A could be said to have such knowledge he could never tell individual B about it. Communication could take place only if A and B could agree that something which is being observed by each of them could be represented by the same symbol. Without symbols the cooperative activities of science would be impossible.

2) Without symbols the permanent and enduring character of science would be lost. Objects could be known only *while* they are being observed; after the collecting process had stopped the data would cease to exist. Even if they were retained for a short time in memory we should have no science, for memory as a kind of symbolism is notoriously unreliable and incapable of retaining more than a small part of what has been directly experienced. Without language the science of any age could not build upon the work of its predecessors.

3) If "knowledge" were restricted to that obtained through direct experience, the theoretical content of science—its hypotheses and conjectures, its explanations in terms of what is "behind" and "beneath" and "inside" apparent objects—would be impossible. Since the consideration of this aspect of science will occupy our

---

[1] A. N. Whitehead, *Symbolism, Its Meaning and Effect* (New York: The Macmillan Company, 1927), p. 6.

attention throughout the remainder of the book we may dismiss it here with a mere mention.

Our task now will be to consider the nature of symbols,[2] how and what they mean, and how they are differentiated into kinds.

1. Signs and Symbols

Languages are systems of individual symbols, and the most convenient approach to the notion of symbol is through the more general concept of *sign*. A sign is always found in a sign-situation. Ogden and Richards[3] point out that if we stand near a crossroad and observe a pedestrian reading a notice *To Grantchester* we suppose there to be three elements in the situation: first, a sign; second, a place referred to by the sign; and, third, a person interpreting the sign. All situations involving the use of signs are of this general character. The sign itself, of course, need not be a printed one, but may be anything capable of being observed, provided it has the property to refer to, or to "signify," something else. The reference need not be to a place, but could be to a time (as in the case of a clock), or to any objects whatsoever in the past, present, or future. We commonly call a sign "natural" when it is itself a natural object and when it calls its referent to mind because of cause and effect relations or other associational relations which are found in nature. For example, a demolished town signifies that a tornado has passed by, congested lungs signify pneumonia, a red sunset signifies a beautiful day to follow, and scratches on rocks signify that a glacier has covered the area. The relation of signifying will be later seen to be the basis for the flash of insight which enables scientists to discover new hypotheses. Contrasted with natural signs are conventional signs. These are called "conventional" either because the signs themselves are artificial rather than natural objects, or because the signi-

[2] Linguistic analysis and the theory of signs have received much attention in recent years. Some references are the following: C. W. Morris, *Signs, Language and Behavior* (New York: Prentice-Hall, Inc., 1940); C. W. Morris, "Foundations of the Theory of Signs," *International Encyclopedia of Unified Science* (Chicago: University of Chicago Press, 1938), I, No. 2; C. K. Ogden and I. A. Richards, *Meaning of Meaning* (New York: Harcourt, Brace & Company, 1930); Stuart Chase, *The Tyranny of Words* (New York: Harcourt, Brace & Company, 1938); and Bertrand Russell, *An Inquiry into Meaning and Truth* (New York: W. W. Norton & Company, 1940).

[3] Ogden and Richards, p. 21.

fying relationship is determined by convention instead of by relations holding in the physical world. Thus a flag at half-mast signifies death of a prominent person, an olive branch signifies peace, a red-white-and-blue striped pole signifies a barbershop, and a red traffic light signifies a stop. The interpreter of a sign need not be a person, since animals behave in such a way as to lead us to believe that they react to signs of certain kinds; for example, the dog who runs to the dining room at the sound of the dinner bell exhibits sign-behavior.

Since scientific symbols are special kinds of signs, they may be expected to exhibit the three aspects mentioned above. In the case of each symbol there is (1) the symbol itself (called the "symbol-token") to which the referential property is attached; (2) the referential property, which somehow "connects" the symbol-token with that which it symbolizes; and (3) the psychological interpretation of this referential property when it enters into the consciousness of the scientist. In order to make these three features clear let us discuss them in greater detail.

1) Every symbol-token is an event, or object, or happening. It may consist of a written word made up of particles of ink arranged in a certain pattern; it may be a noise uttered by a person in an act of communication; it may be a drawing, or photograph, or material model; it may be a group of mathematical representations. A symbol-token must be capable of being observed. This means that it must exhibit sensuous properties and have enough constancy in appearance so that we can recognize it when it occurs. Most symbol-tokens are visual or auditory. Odors, tastes, and touch sensations may function as sign-tokens but do not ordinarily occur as symbol-tokens since they are seldom organized into language systems and rarely used for purposes of communication.

2) Every symbol-token refers to something other than itself.[4] This feature of a symbol-token is complex and involves (*a*) the referential property or relationship, and (*b*) the entity which is referred to. But these two features of the symbol-token do not belong to it in the same sense. The referential property must be "added to" the symbol-token in order that it should have meaning and thus become a symbol. But the *referent* (as contrasted with the *reference*) need not be given in the situation at the moment

[4] This is not strictly true. For example, "word" refers to itself as well as to something other than itself.

except under the form of another symbol. To consider an example, the word "circle" must have a referential character in order to be a symbol, i.e., it must have the power to point to what would be identified as a circle were it given; this is commonly called the "semantical dimension" of the symbol.[5] But no actual circle may be given, though in using the word we may find ourselves thinking of an image of a circle, or of certain words such as "equality of radii," "conic section," "two-dimensional plane figure," etc., and we may believe that these are the referents of the symbol. It is the essential virtue of symbols that we can use them in the absence of their referents, and hence as substitutes for those events and objects to which they refer.

Just *how* symbols refer to events is hard to say. It is convenient to picture the referential property of a symbol by means of an arrow; this suggests in a vague way certain of the features of the reference. It indicates, for example, if the arrow is properly drawn, that the reference somehow takes its origin in the symbol itself as a mere object, and goes out of and away from this source in the direction in which its referent is to be found. It also calls attention to two of the most important features of the reference, viz., its asymmetric character, and its capacity to limit the range of possible events referred to. For if the word "circle" refers to an actual circle, the actual circle cannot refer to the word in the same way. And through the word we are able to know, in advance of seeing the actual circle in question, the *kind* of event to which reference is made. Furthermore, the arrow is a convenient mode of representing the referential property because an arrow may point without pointing to anything, i.e., it may have a general directional indication but when we examine the locality thus called to our attention we may find nothing. Symbols enable us to talk about absent objects, but, unfortunately, by virtue of this same capacity, permit us to talk about objects which are not even known to exist at all. Hence it is an important feature of symbols that we may use them without any prejudice to the existence or non-existence of their referents.

3) Symbols refer only through the intermediary of a mind. Words do not "mean" anything by themselves, though there was a time when this was not so generally recognized. Hence the only way to understand how symbols mean is to ask ourselves what we

[5] Morris, "Foundations of the Theory of Signs," p. 6.

experience when confronted by them. Consider, again, the symbol-token "circle." There is, first, the awareness of the physical object, the word, which is the pattern of ink on paper. This, however, does not usually enter into clear consciousness and may not be distinguishable as an element of the experience. But from this initial experience there arises a consciousness of the meaning of the symbol. This associative tie is a convention basically dependent on past experience; the awareness of the physical symbol calls up the awareness of the meaning through associative ties built up in learning the language. Something is called to mind when we are presented with a word whose meaning we know. The fact that it is called to mind immediately and without conscious effort on our part, leads us to believe that it is somehow a part of the symbol itself. However, the attempt to read in a foreign language soon convinces us to the contrary; here the associations must be made slowly and painfully before they arise with anything like the readiness of meanings in our own tongue.

What we are aware of when we are aware of a meaning is a difficult problem. Eaton considers that the simplest solution is to accept a unique *meaning activity*. "The meaning activity is one of vague anticipation; the mind is poised expectantly, awaiting something other than the thing, the symbol, which is immediately before it; and this anticipation is *vague* because it is not accompanied by a belief that the object meant will appear or that it exists. When I mean an object I do, in some sense, prepare my mind for a presentation of this object. Though I cannot be said to turn my attention toward the thing I mean, since one cannot attend to something not presented to him, there is no doubt that I do more than attend to a symbol or an image. Indeed, I turn my attention away from the symbol or the image, and this constitutes the first step in preparation for the thing meant."[6] The fact that the meaning activity itself is vague creates all the problems associated with the need for definition in language. Since we cannot say clearly what the meaning activity in general is, we cannot be sure in any given case whether a symbol is understood. Symbols may be graded in clarity, ranging from such words as "circle" where the meaning is clearly grasped, to purely nonsense

[6] R. M. Eaton, *Symbolism and Truth* (Cambridge: Harvard University Press, 1925), p. 23. Reprinted by permission of the publisher.

symbols, where the verbal form suggests that there is a meaning yet no meaning is forthcoming. Between these would be found symbols of increasing vagueness, where the meaning is relatively unclear. If we use the arrow to represent the referential character of the symbol in each case, we may say that the word "circle" has a definite arrow which points clearly to a limited area where the referent is to be found; that nonsense symbols, such as "boojum," have no arrows at all, hence are not properly symbols; and that symbols, such as "religion," "justice," and "self," have arrows but point to extensive and vaguely defined areas within which their referents may be found. Definition removes obscurity in two ways—by pointing to situations in which the referent is given for direct awareness, and by indicating relations to other symbols presumably better known.

The distinction between the two terms "sign" and "symbol" has not been stabilized in recent literature. I should like to state my usage. When a sign is natural I shall commonly call it a "sign," without any qualifying adjective. When a sign is conventional, either because it itself is a manufactured object or because its referential relation is determined by the general agreement of its users, I shall call it a "symbol." Symbols are used primarily as instruments of communication. Most actual foreign languages, therefore, and, even the "language of science," are systems of symbols with complicated inter-symbolic relations (sometimes called "syntactical relations").[7] The commonest examples of symbols are words and sentences, images, gestures, drawings and graphs, numbers, mimetic sounds, and physical models. So far as science is concerned we shall be concerned only with such of these symbols as are capable of being interpreted by human beings. (I do not know any dogs who are scientists.) Whether a complex of symbols is or is not a language depends upon several factors, including its scope, the complexity of its syntax, and its use and referential character. For our purposes this distinction is of no concern. The language of science certainly is a language. Whether there are also languages of flowers, of traffic lights, of flags, or of gestures will not be discussed in what follows.

There is one more limitation on our use of the term "symbol." In ordinary language symbols play various roles—cognitive, emo-

---

[7] Morris, "Foundations of the Theory of Signs," p. 7.

tive, directive, interrogative, performative, and many others. Symbols cannot be divided into *kinds* on the basis of their uses since a given symbol may have one use in one context and another in another. Suffice it to say that we shall restrict ourselves to cognitive symbols, unless specific indication is given to the contrary. This does not forbid us, for example, to talk about the emotions, as we must do in behavioral sciences, where the word "happiness" might play a legitimate cognitive role. But it does prevent us from using the word "Hurrah!" since this *expresses* or *evokes* a feeling of happiness and is thus not cognitively used. Symbols in use *exhibit* their "pragmatic dimension." [8]

2. Kinds of Symbols

One of the most important types of symbol occurring in scientific language is the *proper name*.[9] A proper name, in its pure form, is essentially an identifying tag; it serves only to locate an object, not to characterize it or specify its properties. It plays essentially the same role as a gesture of pointing, and is used to call attention to an object and to distinguish it from other objects for which it might be mistaken. It is "attached" to an object when we plan to refer to the object repeatedly and wish to give it an identifying mark. If two objects are likely to be confused with one another we give each a distinct proper name. A proper name is sometimes called an "index"; this reveals its capacity for "locating" its referent.

A pure proper name is distinguished from a common name in two very important ways. In the first place, the former is meaningless if it has no referent. There can be no proper name which is the name of nothing. The correct use of a proper name therefore involves a definite implication of the existence of something which is thus named. There is no such requirement in the case of a common name. We can talk about electrons and perfect gases in a meaningful way even though we are not certain that there are such things. We can also use common names like "fairy" and "winged horse" without implying that there are creatures so designated. A proper name is used only *after* we have become ac-

---

[8] *Ibid.*, p. 7.
[9] Russell, Chaps. VI, VII, and XXIV.

quainted with something which we wish to name; a common name can be used *before* we have observed its referent, and is not meaningless even if there is no such referent.

In the second place, a pure proper name refers to an object by direct pointing; this reference is called its "denotation." A common name refers by specifying the property which is possessed by the object; this is called its "connotation." A common name is said to indicate a predicate, i.e., something which can be attributed to an object. The common name "man" indicates the predicate *human,* and thus refers to a man as an object which exhibits this property. A pure proper name has no connotation but simply denotes. It locates its referent not by means of a property but by means of a tag attached to the object, or by directing the attention of a listener to a particular location in space and time where it may be found.

A possible source of confusion between proper names and common names lies in the fact that the proper names which occur in actual languages are relatively "impure." Names of persons usually indicate sex, and sometimes indicate nationality; names of cities, such as "New York City" or "Eastborough," indicate that they are names of cities or towns rather than of persons or rivers; names of certain individuals, such as "Napoleon" and "Socrates," while purely denotative in origin, have now become connotative and one may legitimately speak of "a Napoleon" or "a modern Socrates." The proper name "Homer" seems to violate the condition laid down above, i.e., the need for a referent, since we do not know strictly that there ever was such a person as Homer. But we can see that "Homer" is not a pure proper name; it is rather an abbreviated description equivalent to "the person who is presumed to be the author of certain poems."

The purest proper names which our actual language reveals are usually not called by this term. They are such words as "this," "that," "here," "now," and "I." The fact that some of these are called "demonstrative pronouns" and others "personal pronouns" indicates that they are substitutes for common names and symbolize their referents by demonstration or pointing rather than by connotation. Even these, of course, are not strictly pure. The fact that we have the contrasting words "this" and "that" shows that they are not purely denotative; the former indicates nearness in space or time, and the latter indicates remoteness, both of which may be called properties of the things referred to. Similarly "here" and "now" indicate spatiality and temporality, both of which are attributes. One interesting characteristic of the relatively pure proper names is that they tend to mean something different in every situation in which they are used. This is quite obvious in the case of the pronoun "I," which varies according to the speaker, and may vary even more significantly since a person who uses it repeatedly is presumably himself changing each moment and the meaning of the pronoun must change accordingly. Some of the complications involved in the use of proper names can be illustrated by the following hypothetical incident: John meets George in the dark and, not being able to recognize him, calls out, "Is that you?" George answers, "No."

Can proper names possess syntactical relations, i.e., relations to other proper names within a larger system of such names? An affirmative answer to this question would seem to be justified in view of the fact that there actually are such systems in the numerals, and in the alphabet, both in their "normal" arrangements. Such systems could then be called "directories," "catalogues," "inventories," and "indexes." Obviously the use of number systems as proper names is playing a rapidly increasing role in our lives. One has only to list social security numbers, employee numbers, student numbers in universities, convict numbers in prisons, ZIP codes, and even telephone numbers. Whether these are pure proper names in the sense of merely identifying objects which bear the numbers or "impure" proper names which partially characterize the objects named varies in individual cases. The numbers on the backs of football players are sometimes pure and sometimes impure, since they often indicate position played. ZIP numbers are certainly not pure proper names, because they are designed to indicate localities. Probably most identifying numbers of objects such as are found in a wholesale catalogue, tell not only where the object is to be found in the warehouse but also generally what *kind* of object it is. I was invited once to visit a friend in Prague, Czechoslovakia. He accompanied his invitation with a mention of his street and house number, and with a map showing the position of the house on the street. This latter item, he insisted, was quite necessary because houses were numbered in that area not by their position on the street but by the order in which they were built. Such a number system was not only impure but (from the point of view of convention) misleading. The question of ordered proper names will come up for reconsideration in a later chapter.

Proper names play two very important roles in science. The first is a temporary one; the second is a more or less permanent one. The former is based upon the obvious fact that all science (except, possibly, pure mathematics) begins with observed data of one kind or another. Let us suppose that a botanist is about to test the effectiveness of several fertilizers. He gathers, perhaps, ten samples of a certain plant and fertilizes each with a different product. Unless he records which plant is treated by which fertilizer he may be able to draw no conclusions. Hence he labels each plant, possibly by attaching a tag bearing the number "1," the number "2," and so on, until all plants have been numbered. Then he labels his fertilizers, possibly A,B,C, . . . and correlates plant 1 with fertilizer A, plant 2 with fertilizer B, and so on until he has exhausted his supply of plants. *If* each fertilizer affects its plant in the same general way the botanist can summarize his results by saying that the *kind* of plant he has been studying is affected in a characteristic way by all of the fertilizers used. This conclusion he can express as an empirical generalization, or, at least, as a summary of cases, and he can destroy the individual labels, 1,2,3,4. . . . They have played their temporary role and are no longer needed. As the number of generalizations increases, the need for proper names decreases. On the other hand, if no generalization is possible because certain of the fertilizers affect their plants in different ways, it is important to the scientist that he be able to correlate the individual plants with the proper fertilizers. The appropriate tags permit this, and they are therefore retained, at least temporarily. Continued reference to uniqueness is necessary until generalization is possible.

The second role of proper names in science is quite different. It presupposes the first role of these symbols, since, as was stated above, all science (again, with the possible exception of pure mathematics) begins with observed, individual objects and events. But certain kinds of science *retain* the initially assigned proper names, since their main task is to provide information about the unique features of individual objects and events, rather than about kinds of events and about general laws stating correlations between classes of events. These sciences are commonly called "ideographic," while the latter are usually characterized by the term "nomothetic."[10]

> It is often maintained that the natural sciences and some of the social ones are nomothetic, whereas history (in the sense of an account of human events, as distinct from the events themselves) is pre-eminently ideographic. In consequence, it is frequently claimed that the logical structure of the concepts and explanations required in human history is fundamentally different from the logical structure of concepts and explanations in the natural (and other "generalizing") sciences. . . . Even a hasty inspection of treatises in theoretical natural and social sciences on the one hand (such as optics and economics) and of books on history on the other hand, suffices to reveal a striking difference between them. For by and large the statements occurring in the former are general in form and contain few if any references to specific objects, dates or places, whereas statements in the latter are almost without exception singular in form and replete with proper names, designations for particular times or periods, and geographic specifications. To this extent, at any rate, the contrast between the natural and some of the social sciences as nomothetic, and human history as ideographic, appears to be well founded.[11]

Nagel's inclusion of history among the sciences need not disturb us, since we can say simply that history, if it *were* a science, would be an outstanding example of an ideographic science.

We can see, then, that even if we limit our discussion to the natural and behavioral sciences, the decision with regard to whether one of these sciences is dominantly ideographic or dominantly nomothetic is made on the basis of the preponderance or the infrequency of the occurrence of proper names in that science. Geography is one of the most obvious of the ideographic sciences, since the geographer would be literally speechless if he were not permitted to talk about individual lakes, rivers, mountains, cities, countries, oceans, and other physiographical phenomena and social groupings capable of being located by latitude and longitude on the surface of the earth. The contrast between the two types of science is well illustrated by astronomy and celestial mechanics; in the former there are frequent references to individual stars,

---

[10] These terms were introduced by Wilhelm Windelband in his essay, "*Geschichte und Naturwissenschaft*," reprinted in the collection of his essays, *Präludien* (5th ed., Tubingen: J. C. B. Mohr, 1915), II, 136–160.

[11] Ernest Nagel, *The Structure of Science* (New York: Harcourt, Brace & World, Inc., 1961), pp. 547–548. Reprinted by permission of the publisher.

planets, asteroids, comets, and even to our own sphere, while the latter concerns itself mainly with the laws of the movements of various kinds of heavenly bodies. For example, astronomy tells us about the moon (the expression "the moon" functions as a proper name), its distance from the earth, its time of revolution about the earth, its own rotation, its size and shape, and so on; on the other hand, Kepler formulated the laws of planetary motion, which, while they included reference to the sun, did not contain explicit reference to any of the known planets and presumably will apply to any new planets which may be discovered in the future. Geology is strongly ideographic, including in its vocabulary names of mountains, rivers, glaciers, valleys, mesas, volcanos, and oceans. Anthropology and archaeology are also dominantly ideographic. Most of the social sciences are partly ideographic and partly nomothetic. For example, a sociologist might make a study of the city of Chicago—the racial distribution of its population, its housing, its transportation, its industry, its churches and schools; such an investigation would be ideographic. On the other hand, a sociologist might inquire into the laws of movements of large masses of people, from urban to rural areas or the reverse, from backward countries to countries with high standards of living, from industrial areas in cities to residential areas, and so on. Similarly a political scientist might make an analysis of the structure of the present American democracy, or he might pursue an investigation of the features which are common to all democracies. At the other end of the series would lie the dominantly nomothetic sciences, such as physics, chemistry, botany, zoology, and physiology. These sciences would use proper names very infrequently.

Our later considerations will show that there could be no purely ideographic sciences, since even in history, where the major concern is with particular events in chronological sequence, the use of sociological generalizations attempting to explain *why* a given succession occurred would be required if history is not to be mere chronology. No statement concerning the properties of individual events and objects can be made without the use of generalizations of some kind. Correspondingly, there could be no purely nomothetic sciences, since no information can be obtained about an individual object, i.e., anything in the world, unless in addition to our knowledge of the laws of objects, we know through observation that there are objects exemplifying the laws.

Another very important type of symbol occurring in science is called by C. S. Peirce an "icon." This is defined by him as any symbol which "may represent its object mainly by its similarity." Illustrations of icons are to be found in *images* "which partake of simple qualities"; *diagrams* "which represent relations, mainly dyadic, or so regarded, of the parts of one thing by the analogous relations in their own parts";[12] and *metaphors* which represent through parallelism. "It is a familiar fact that there are such representatives as icons. Every picture (however conventional its method) is essentially a representation of this kind. So is every diagram, even though there be no sensuous resemblance between it and its object, but only an analogy between the relations of the parts of each."[13] Icons can be actual physical models, such as are found in a museum—models of the solar system; of the atom; of the human body made with a transparent "skin" and showing all internal organs variously colored; of the chromosome; of animals and flowers in their native habitats; of tribes of people, usually greatly reduced in size, engaged in their daily life of fishing, hunting, cooking, and other social activities; and many others. Geographical maps are one of the most common form of icon: Richard Edes Harrison[14] has devised an effective method for mapping topography by imaginatively placing the viewer high in a plane above the terrain—not directly above, but looking down obliquely so that he can see the curvature of the earth, the mountains and valleys as they are indicated by the sun's shadows, the blue sea, and the green or brown plains. Often these maps are so devised that the observer is not situated so that north is at the top, as is customary in map representation, but is viewing an area such as Japan from a plane flying high over Alaska. Animated cartoons can cleverly depict many of the abstract, scientific properties of objects. Graphs on coordinate systems constitute a type of icon which is not closely pictorial but does have a type of structural similarity to that which it represents.

Icons play various roles in science, and no statement can be

---

[12] C. S. Peirce, *Collected Papers*, ed. C. Hartshorne and P. Weiss (Cambridge: The Belknap Press of Harvard University Press, 1932), II, par. 2.276 and par. 2.277.

[13] *Ibid.*, par. 2.279.

[14] Richard Edes Harrison, *Look at the World* (New York: Alfred A. Knopf, Inc., 1944).

made about their general effectiveness as portrayal devices unless one considers the special role they are playing. As popularizers of science, through the various communication media—television programs, popular science magazines, newspapers, and museum displays—their virtues are many. There are, however, risks involved. No icon is a perfect representation; it is always in some sense an over-simplification or it introduces elements into the icon which are not in the object portrayed. Picturing is most often obtained by changing the size of the object depicted, by stopping or changing the rate of its motion, by transforming it from a solid into a two-dimensional figure, or by rendering concrete something which is essentially abstract, as in the case of the portrayal of time as a flowing river or an unrolling carpet. Furthermore iconic representation is difficult, if not impossible, in most of the social sciences (unless we take them in their ideographic form), for we cannot make a picture of democracy, honesty, egoism, patriotism, extroversion, or internationalism.

The variety of roles played by icons in science has suggested to certain writers that there may be among scientists two basic types of mind—one which prefers concrete imagery, models, and pictures, and the other which favors abstract representations, such as are found in language, especially in a mathematical language. Certain English physicists, among whom were Faraday, Lord Kelvin, Lodge, and Maxwell, argued strongly for the virtues of concrete representation as over against more abstract forms of portrayal. Kelvin, for example, said in effect that the question, "Do we understand a certain physical subject or not?" means "Can we construct a corresponding model?"[15] "I advise you all who are engaged in teaching, or in thinking of these things for yourselves, to make little models."[16] Lodge put this plan into practice when he explained modern theories of electricity by diagrams of machines involving pulleys and cords, weights and drums.[17] As long as one limits oneself, he insisted, to the abstract schemes of mathematicians who are "able to live wholly among symbols, dispensing with pictorial images and such adventitious aids"[18] he does not

[15] W. T. Kelvin, *Baltimore Lectures on Molecular Dynamics* (Baltimore: Johns Hopkins Press, 1884), p. 131.

[16] *Ibid.*, 2d ed., 1904, p. 34.

[17] Oliver Lodge, *Modern Views of Electricity* (2d ed.; New York: The Macmillan Company, 1892), *passim.*

[18] *Ibid.*, p. 13.

truly understand the real phenomena. On the other hand, there is a large group of scientists, among whom are Hobson, Mach, Maxwell, Poincaré, who believe that images and models are apt to be misleading and should be replaced by abstract conceptual schemes or word-representations.[19] Here would be found those who consider the language of mathematics to be *the* language of science.[20]

Duhem has gone so far as to suggest names (following Pascal) for these two extreme types of mind, each so useful in its own way in carrying on science. The mind which favors models is called "*l'esprit de finesse*," for which Duhem's translator uses as English equivalents "the ample mind" and "the supple mind." The mind which prefers the more abstract mode of representation (Pascal's term is "*l'esprit géométrique*") is translated as "the logical mind," "the rigorous mind," and "the geometrical mind."[21] (Duhem carries this distinction beyond science into literature and philosophy.) Obviously the notion of "pictorial representation" can be broadened to the point of absurdity. There is a wide sense in which one can speak of any language as being iconic. Onomatopoetic words, such as "ding-dong" and "bow-wow," are clearly phonetic representations of their referents, Egyptian hieroglyphics are pictorial, and philosophers who talk about the "correspondence theory of truth" are implying a broad iconic relation between language and the world. Wittgenstein's statement, "The proposition is a picture of reality," is in accord with this view.[22]

I have chosen the term "characterizing symbol" to designate those elements of the language of science which refer to the world connotatively rather than primarily denotatively. Thus character-

---

[19] E. W. Hobson, *The Domain of Natural Science* (New York: The Macmillan Company, 1923), Chap. II; Ernst Mach, *The Science of Mechanics* (LaSalle: Open Court Publishing Co., 1907); J. C. Maxwell, *A Treatise on Electricity and Magnetism* (New York: Oxford University Press, 1873); P. W. Bridgman, *The Logic of Modern Physics* (New York: The Macmillan Company, 1927), pp. 52–60; and W. H. Watson, "On Methods of Representation," *Philosophy of Science*, ed. Arthur Danto and Sidney Morgenbesser (New York: Meridian Books, Inc., 1960), pp. 226–252.

[20] This would presumably be the position of J. G. Kemeny as indicated by the "diagram" of science given in his *A Philosopher Looks at Science* (Princeton: D. Van Nostrand Company, Inc., 1959), p. 86.

[21] Pierre Duhem, *The Aim and Structure of Physical Theory*, trans. Philip P. Wiener (Princeton: Princeton University Press, 1954). Chap. IV.

[22] L. Wittgenstein, *Tractatus Logico-Philosophicus* (New York: Harcourt, Brace & Company, Inc., 1922), prop. 4.021.

izing symbols are contrasted, so far as this can be done at all, with pure proper names. They serve in science not merely as labels or tags but provide information about the properties and relations of objects and events in the world. Characterizing symbols are further divided, again on a principle which is largely relative, into iconic and non-iconic. Iconic symbols are more, rather than less, pictorial, and include models, charts, maps, diagrams, pictures, graphs, and other (usually non-verbal) modes of representation. Non-iconic characterizing symbols are pictorial only in a forced sense, and contain the usual nouns, pronouns, verbs, adjectives, adverbs, and sentences of our languages. They include both verbal and mathematical symbols[23] for observed data and for hypotheses; examples are *iron, gas, electricity, mass, electron, proton, organism, plant, animal, cell, gene, chromosome, human behavior, family, government, commerce, money, stimulus,* and *response.* Characterizing symbols also include all laws, universal or statistical, which state correlations among the observed and hypothetical objects and events just listed—between observed objects and observed objects, between observed objects and hypothetical objects, and between hypothetical objects and hypothetical objects.

Finally, in order to complete our discussion of the main types of symbols found in every scientific language we must mention two kinds, whose role in science cannot be examined until we have progressed further with our analysis. The first of these is called *logical*[24] symbols, or, sometimes, *structural*[25] symbols. Only after we have seen something of the pattern of the scientific method can we understand the way in which these symbols enter into science. They include such logical words as *premise, implies, either, not,* and such methodological words as *observation, description, explanation, confirmation, induction,* and *experiment.* The second type of symbols constitute *meta-symbols.* They are, as the term indicates, symbols which refer to symbols. For example, if I say that the word "cat" has three letters I am predicating something not of a cat but of the word "cat," and the expression "has three letters" is meta-linguistic. We shall see in the course of our further study

[23] Peirce, par. 2.279. Peirce here argues that mathematical symbols are icons.
[24] Russell, Chap. V.
[25] Israel Scheffler, *The Anatomy of Inquiry* (New York: Alfred A. Knopf, Inc., 1963), p. 5.

that since science contains symbols, if we are to talk about these symbols we must use meta-symbols. Indeed, without meta-symbols this book could not have been written.

The main types of symbols occurring in a scientific language can be summarized in Figure 3.

FIGURE 3

THE LANGUAGE OF SCIENCE.

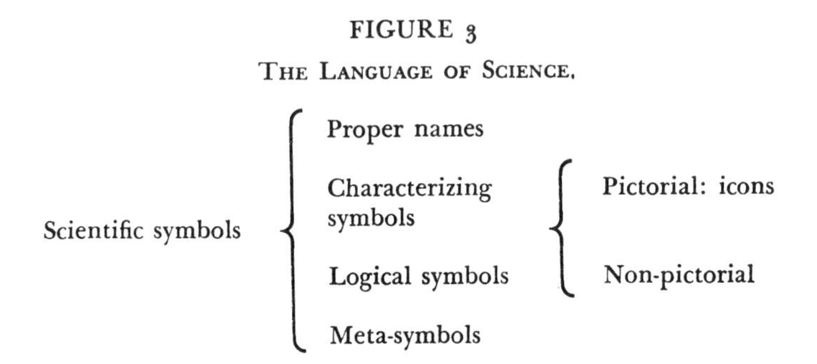

Describing the Facts

We now have the tools with which to proceed in our analysis of the scientific method. At the beginning of the preceding chapter we discovered that a mass of facts does not constitute a science; the facts must be described and explained by means of a language adapted to play that role. The general features of this language have now been examined.

The attempt to distinguish sharply between *description* and *explanation* will be postponed until the next chapter. For the present we shall use these words as they are employed in ordinary speech. When we describe an object or event we usually attempt to symbolize its more obvious and apparent properties—the so-called actual properties disclosed through direct observation, and some of the dispositional properties obtainable both by performing simple transforming operations and observing the results of these operations, and by introducing recording instruments which do not actually modify the material we are examining but merely multiply its manifestations to the senses and thus increase our knowledge of the properties it possesses. For convenience of presentation we shall divide the descriptive operations performed in obtaining this information into four groups: classifying operations, associating operations, ordering operations, and measuring operations.

1. Classifying

Classification[1] is the process of putting objects or events into classes by virtue of properties which they possess in common. Robins, for example, are all birds with gray backs and reddish brown breasts, about nine inches in length, commonly appearing in the temperate zones early in the spring, and feeding mainly on worms. Of the descriptive operations listed above, classification is most closely identified with "description" as used by the man in the street. Classes are symbolized by such characterizing symbols as *concepts, class-names, adjectives,* and *common nouns.* Classes are logical entities and should be distinguished from groups or collections. The members of a class need not be spatially or temporally united, as they are in a collection, but are joined by their possession of a common property. Thus a forest would be distinct from *the class* of trees, and a herd of cows should not be confused with *the class* of cows. Sometimes members of the same class are grouped together and put into boxes or other compartments. This operation may happen in science in its preliminary stages, and commonly occurs outside science in the case of museums, libraries, post offices, department stores, cities which are zoned for industry and for residence, and in many other types of situation.

When a common name is already available for use, as in the case of the word "robin," classification is mere identification or recognition. Often in science, however, our common language contains no word to portray a given property; in such cases new words must be invented. This is common in the naming of new chemical elements and compounds, new plants and animals, new types of national government, new types of mental disorder, new forms of social organization, and new types of business transaction.

[1] For more detailed discussion of *classification* see the following: W. S. Jevons, *Principles of Science* (London: Macmillan & Co., Ltd., 1907), Chap. XXX; H. A. Larrabee, *Reliable Knowledge* (Houghton Mifflin Company, 1945), pp. 239–248; Jerome Bruner, *A Study of Thinking* (New York: John Wiley & Sons, Inc., 1956); and R. W. Church, *An Analysis of Resemblance* (New York: The Macmillan Company, 1953).

Classes are well defined to the extent to which the terms used to characterize them are themselves precise in meaning. A very young child has difficulty distinguishing dogs from cats, and even an adult, unless well informed, may be unable to distinguish dogs from wolves.

The preliminary classification of objects is often based on superficial resemblances and differences, and these often need to be changed as the science reaches a higher level of development. The whale is commonly classified with the fish, which it resembles in general habitat and mode of living. But the whale possesses milk glands and suckles its young, and is therefore more fruitfully classified with the other mammals. Sugar and saccharin are clearly similar in their white color and sweet taste. But chemically they are quite different, for sugar is a carbohydrate containing carbon, hydrogen, and oxygen, while saccharin has the chemical characteristics of an acid, and its molecule contains carbon, hydrogen, oxygen, nitrogen, and sulphur. The naive astronomer classifies all of the "stars" together, since they emit light at night; but some of them are self-luminous and some are not, and some of them are fixed in position among the other neighboring heavenly bodies, while others are in motion. A bat looks and behaves like a bird; yet a bat is a mammal while a bird is not. The amateur geologist thinks of all "hills" as alike; yet some have been thrust up from below, others are the remains of mountains which have been eroded, and still others are formed by glacial deposit, by volcanic eruption, by avalanches, and by other means.

Negative examples support this same claim. The average individual would not be inclined to place the potato plant and the tomato plant in the same class, or the yellow buttercup and the hooded blue aconite; yet the botanist, who is not confused by apparent differences, places these together. In their natural states chlorine is a gas, bromine is a liquid, and iodine is a solid; but they resemble one another closely in chemical properties. Light and electricity seem to be quite different; yet they are examples of the same phenomenon. In all these cases apparent differences, revealed through direct observation, obscure basic resemblances. In the language of Aristotle, no longer accepted by scientists today, the superficial similarities are "accidental," while the deeper similarities are "natural" to the objects concerned.

Following Larrabee[2] we might better call such preliminary classifications, "diagnostic" or "exploratory" rather than "artificial."

Not only may objects be united into classes by exhibiting common properties, so may classes be united into larger classes in the same way. This operation gives rise to a system of classification, such as given in Figure 4. The common song sparrow is placed in the class of other song sparrows which it closely resembles; and this class is then subsumed, on the basis of more general resemblances, in a larger class, and so on, until the phylum is reached, where the resemblance is merely the possession of a notochord (a longitudinal elastic rod of cells) which forms the axis of the body.

FIGURE 4

BIOLOGICAL CLASSIFICATION OF THE SONG SPARROW

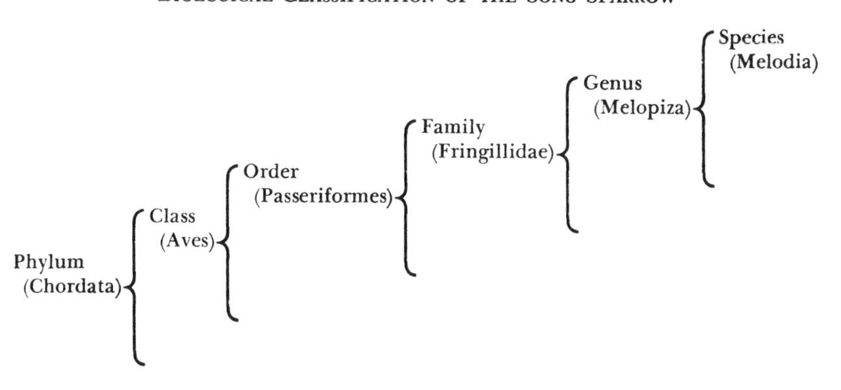

It can be readily seen that a classificatory scheme has a general pattern.

FIGURE 5

GENERAL CLASSIFICATORY SYSTEM

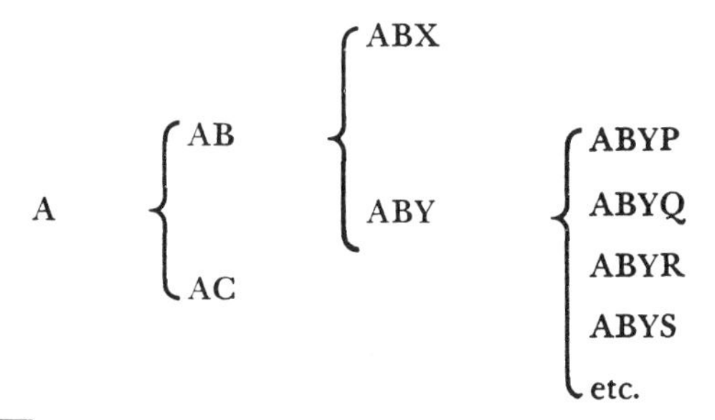

[2] Larrabee, pp. 242–243.

A is called the "*summum genus,*" and represents the most inclusive class—the class which has been obtained from the originally observed objects by the highest degree of abstraction. Other classes, proceeding to the right, are sub-classes and sub-sub-classes of the *genus* until the original class of objects is reached. Proceeding in this direction we are said to subdivide instead of to classify, and the movement is from abstract to concrete rather than the reverse. This relation of classes will be of great importance when we come to examine the mathematical sciences, which are commonly considered to be the most highly abstract of all sciences. Rules have been established by logicians for setting up ideal classificatory schemes. Of these, three are usually mentioned: (*a*) the division of a class into sub-classes must be exhaustive, i.e., every member of the larger class must lie in one of the sub-classes; (*b*) the sub-classes into which a class is divided must exclude one another, i.e., there must be no overlapping of classes; and (*c*) so far as possible a single principle of subdivision must run throughout the entire system. These rules are so simple as to require no further comment. Their actual application, however, is often extremely difficult in the pursuit of science itself.

Darwin was of the opinion that classification was the very essence of science. "Performed consciously or unconsciously, the act of classification is indispensable to and accompanies every scientific inference. A mind is orderly or slovenly, according as it does or does not habitually and accurately classify the facts with which it comes into contact. The success of an investigation, the worth of a conclusion, are in direct proportion to the fidelity to this principle and the exhaustiveness with which the process is carried out."[3]

2. Associating

A second descriptive technique is *association*. In one sense this depends on classification. In fact it involves nothing more than the attempt to determine how frequently certain objects which we have examined fall into two classes at once. How often have

---

[3] Frank Cramer, *The Method of Darwin* (Chicago: A. C. McClurg & Company, 1896), p. 88. Reprinted by permission of the publisher.

the metals which we have examined proved to be also conductors of electricity? How often have we discovered that people who develop lung cancer are also heavy cigarette smokers? Of the births recorded in a certain hospital how many have been males and how many females? The answers to these questions are commonly mistaken for *scientific laws*, either universal or statistical. But to characterize them in this way is confusing. The important property of a law is that it claims to "go beyond" observed cases and make predictions as to what kinds of instances will be found in the future. On the strictly descriptive level this is impossible. For example, in answer to the above questions we might find that all the metals *we have examined* were also conductors of electricity, that of the 125 people with lung cancer treated by a certain doctor 80 were heavy smokers, and that of the 242 births recorded in a certain hospital 124 were male and 118 were female. These may *suggest* the laws that all metals are conductors of electricity, that when large numbers of people with lung cancer are examined it will be discovered that roughly 64 per cent of them are heavy smokers, and that in the population at large the ratio of male births to total births is approximately 51 per cent. Such laws are *hypotheses*, to be confirmed or disconfirmed by later observation, and are not found at the strictly descriptive level. Furthermore, to go beyond the hypothesis and to assert causal connections (smoking is the cause of lung cancer) would be even less justifiable at this stage of investigation. Correlations in descriptive science are restricted to what are sometimes called "enumerative generalizations" or "statements of observed frequencies." The former vary from such trivial cases as "All the months of the year contain more than twenty-five days" and "All of the Apostles were Jews" to cases expressed by "All known gases expand when heated" and "All known non-albino crows are black," where observed exceptions have been taken care of by showing that in each case there was a "disturbing factor" present. Statements of observed frequencies are illustrated by the life-expectancy tables commonly used by insurance companies, indicating how many of the people who have died in a given period succeeded in reaching the age of sixty, how many of them were day laborers, and how many were professional men, how many died of tuberculosis, or heart disease, or cancer, and so on.

3. Ordering

Ordering operations are descriptive of the serial relations of objects and events. For our purposes the operation of ordering will be greatly simplified, and we shall introduce certain restrictions for the purpose of making our exposition as clear as possible. First, we shall confine ourselves to classes where the members are alike in some designated respect. Second, we shall presuppose that we have a class of at least three members. No doubt a class of two members (x,o) can be ordered either as x is to the left of o or as o is to the left of x, and these would clearly be different orders (or at least different *arrangements*). But they would not be significantly different because each arrangement could be changed into the other merely by looking at one through a mirror. On the other hand, a class of three members (a,b,c) could be changed into the order (a,c,b) or the order (b,a,c) and by no instrumental manipulation or change in the position of the observer could the different orders be made alike. Third, we shall make the same distinction which we did in the cases of classes: classes are logical entities, and collections are physical groupings; similarly, series (or orders) are logical entities, and arrangements in space or time (as of books on a shelf or objects in a museum display) are physical placements which presuppose that the objects or events have been brought together in some locality or temporal period. Thus the descriptive operation of ordering is, like the descriptive operation of classifying, mainly a "mental" or symbolic operation rather than a physical one.

I shall proceed first to give a technical definition of "ordering"[4] and then to present examples of ordered groups which occur frequently in science.

A class is ordered by means of an *ordering relation.* A dyadic (two-termed) relation is an ordering relation with reference to a class if it is (1) asymmetrical, (2) transitive, and (3) connected.

A relation is (1) asymmetrical when if it holds between x and y it cannot also hold between y and x. Thus an asymmetrical relation is one in which the order of terms is significant, i.e., one in

[4] A simple linear order, such as we are discussing here, may be defined in alternative ways. I have chosen the postulates given by Bertrand Russell, *Introduction to Mathematical Philosophy* (New York: The Macmillan Company, 1919), Chap. IV. An alternative system is given by E. V. Huntington, *The Continuum and Other Types of Serial Order* (Cambridge: Harvard University Press, 1917), Chap. II.

which an interchange of the terms without a change in the relation would produce a relation-complex incompatible with the original. For example, if a substance a scratches b, then b cannot scratch a; or if x is the father of y, then y cannot be the father of x; or if p is longer than q, then q cannot be longer than p. But if $k=l$, then $l=k$; hence equality is symmetrical.

A relation is (2) transitive if when it holds between a first term and a second term, and between a second term and a third term, it then holds between the first term and the third term. For example, if substance a scratches b and b scratches c, then a scratches c; and if p is longer than q and q is longer than r, then p is longer than r. But if x is the father of y and y is the father of z, x is *not* the father of z but his grandfather.

A relation is (3) connected with reference to a class if it holds between every pair of elements in the class. For example, we could not, in general, order a class of straight lines by the relation "shorter than," since there might be two lines of exactly the same length; but we could do this in the case of a class of lines of different length. We could order the tones on a piano by the relation *lower in pitch than,* since there there are no two notes of the same pitch; but we could not do this in the case of an orchestra since the tone produced by an oboe might be the same in pitch as that produced by a violin. The class of lineal descendants of a certain man could be ordered by the relation *lineal descendant of,* but the class of *all* his descendants could not be so arranged for two of his descendants who are brothers would not have the required relation. The inclusion of the property of connectedness in the definition of "order" is merely a matter of simplification for the purposes of this presentation. It means that when we turn to those forms of order which are quantitative[5] we shall restrict ourselves to classes of elements no two of which have the *same* quantitative value.

These three properties defining "order" are independent of one another. A relation may be asymmetrical but intransitive (*one inch taller than*), transitive but symmetrical (*the same color as*), asymmetrical but not connected with reference to a certain class (*one year older than* as applied to the children of a certain couple), or transitive but not connected with reference to a certain class (*longer than* as applied to a class of lapses of time).

[5] See below, p. 72 and footnotes.

When the relational property exhibits all of the defining properties mentioned above, the class may be ordered. This is done by arbitrarily deciding, say, that if *x* bears the relation in question to *y* it shall be written to the left of *y*. Then the asymmetry of the relation prevents *y* ever occurring to the left of *x*. The transitivity of the relation means that if *x* has the relation to *y* and *y* has it to *z*, then the order *x,y,z,* alone will be possible. The connectedness of the relation guarantees that every element of the class will find one and only one position in the series. For example, hard objects may be arranged in a series, beginning with *a* which scratches all other members and ending with, say *k*, which is capable of being scratched by all other members, and with the remaining members being properly placed between *a* and *k*.

In order to condense the illustrations of orders which are commonly found in science I have drawn up the following chart:

FIGURE 6

ORDERED CLASSES

CLASS	ORDERING RELATION	LINEAR SERIES
Presidents of United States	Precedes in office	Washington, Adams, Jefferson, Madison, Monroe
Geological ages	Precedes in time	Archaeozoic, Proterozoic, Paleozoic, Mesozoic
Events of the 19th century	Includes in time	Lincoln's life, Civil War, Battle of Bull Run
Organic units	Spatially includes	multicellular organism, cell, chromosome, gene
Social units	Includes	state, county, community, family, individual
Planets	Farther from the Sun than	Pluto, Neptune, Uranus, Saturn, Jupiter
Governmental units	Includes "structurally"	U.S. government, legislative branch, Senate, Senator
Geometrical figures	More abstract than	plane figure, rectilinear plane figure, triangle, equilateral triangle
Points on the Equator	West of	10° E. longitude, 20° E. longitude, 30° E. longitude . . . up to 180° E. longitude
Causal events	Causes	striking of match, ignition of match, ignition of fuse, explosion of gun powder

It may have been noted that some of the series considered above were defined by relational properties which introduced comparative terms, and some by properties which did not. For example, if the lineal descendants of a man are arranged serially there is no sense in which we can say that one person in the series is any more or less of a descendant than another; but if we arrange lengths by the relation *extends beyond* we can say that any length which *extends far beyond* another is longer than one which *extends only slightly beyond* the other. Ordering relations which permit this interpretation may be called "comparative" and such series may be called "comparative series." They arise when, in our effort to go beyond the crude discriminations of common sense, we attempt to introduce greater precision by replacing differences in kind by differences in degree. Classificatory systems are based upon differences in kind. Membership in a class is determined by the possession of a certain quality; everything in the world which possesss the quality lies within the class and everything which lacks the quality lies outside the class; there are no objects which lie more in than out. Classificatory systems are "either-or," not "more-or-less." Comparative notions occur when the "opposite" ideas of *presence* and *absence* are replaced by the relative ideas of *degrees of presence* or *of absence*; or when contradictory terms, such as "tall" and "non-tall," are replaced by contrary terms, "tall" and "short," with an intervening "medium height." Thus properties are comparative only because they hold in at least a *three-term situation,* where the comparative relation of one element to a second is the same as the comparative relation of the second to the third.

We may therefore say that a class can be arranged into a comparative series if there exists an ordering relation which is capable of being interpreted as *more of something than.* It will be noted that this relation has the required properties of asymmetry, transitivity, and connectedness (if each of the elements is of different degree). All of the usual quantities of physical science (space, time, mass, conductivity, elasticity, resistance, density, valence, solubility, metabolic rate, osmotic rate, rate of erosion or of weathering or of ageing) can be ordered on a comparative basis. In addition, many classes which are peculiar to the psychological, the humanistic, and the valuational sciences can be roughly ordered comparatively. We can order tones on the basis of their heard intensities or light sensations on the basis of their seen intensities. We can order individuals on the basis of the relation *more intelligent than,* or the relation *more loyal than,* or the relation *more gregarious than.* In a very crude sense we can also arrange aesthetic objects in terms of their beauty, moral virtues in terms of their nobility, and religious acts and attitudes in terms of their piety.

Whether comparative series can actually be established in each of these cases is to be determined empirically. The setting up of such orders appears to be increasingly difficult as we pass from the natural sciences to the behavioral studies. If we take aesthetic values as a typical example we can see something of the difficulties involved. Certainly we should have to restrict ourselves to a specific artistic form, such as music, for it would be absurd to ask whether a symphony is more beautiful than a painting. But can we compare instrumental music with vocal music on a single dimension? Or if we limit ourselves to instrumental music can we compare solo work with ensemble work? And can we compare composers or only the individual works of composers? And do we compare the works or only the playing of the works? And do we compare the simple playing of the work or the playing as it fits into our mood at the time of listening?

If we grant the establishment of comparative orders, certain questions of terminology arise. Should comparative series be called "quantitative"? Usage has not become standardized on this matter. If the term "quantity" is restricted to that which is capable of measurement, series of this type should not be called "quantitative," for some of those mentioned above permit measurement and others do not. Apparently, therefore, something in addition to the simple fact of comparative order is required for measurement. On the other hand, the term "quantitative" may not be restricted to that which is measurable. This usage, which will be adopted in the following discussion, permits us to speak of quantities which are not measurable and quantities which are measurable. We may then say that a class of objects can be arranged into a quantitative series if there exists an ordering relation on the basis of which it may be arranged into a comparative order, i.e., the relation *more of something than.* Larrabee gives an interesting list of objects arrangeable into series on the basis of this ordering relation: "oceans by smoothness, razor blades by sharpness, pencils by hardness, hands by cleanliness, paints by durability, foods by appetizingness, drunks by intoxicatedness, teachers by interestingness, and so on."[6] Quantity, therefore, presupposes quality; for *membership* in the class is determined by possession of the common characteristic, but *position* in the serial arrangement of the class is determined by the degree to which the common property is exhibited by each individual member.

4. Measuring

Measurement is the process of assigning numbers to represent qualities.[7] On the basis of the rules for assigning the numbers each number becomes "attached" to the object whose quality it measures, and becomes its measured value. The restrictive condition "to represent qualities" must be introduced in order to distinguish measurement from the mere assigning of numerals as proper names. As we saw in our earlier discussion, this use of numerals is based on no property of the object named, but is arbitrary and the numeral cannot be said to characterize the object.

Yet in spite of the value of measurement as a tool for scientific explanation, its use in the field is not undisputed. Probably no one would argue against its employment in the physical sciences, since the world seems to have been constructed in such a way as to permit its properties to be expressed readily in mathematical language. But at the level of the social and valuational sciences, metrical methods appear to introduce a certain element of artificiality. In these areas comparative series of a very crude sort are sometimes possible, but no measurements in the strict sense. Even where statistical methods are used, the dangers are so many ("You can prove anything by figures," "Figures don't lie but liars do figure") that their employment requires the greatest caution. For

---

[6] Larrabee, p. 376.

[7] C. G. Hempel, "Fundamentals of Concept Formation in Empirical Science," *International Encyclopedia of Unified Science,* II, No. 7, Sec. III; Ernest Nagel, "Measurement," *Philosophy of Science,* ed. Arthur Danto and Sidney Morgenbesser (New York: Meridian Books, Inc., 1960), pp. 121–140; Norman Campbell, *What is Science?* (New York: Dover Publications, 1921), Chap. VI; Norman Campbell, *Foundations of Science* (New York: Dover Publications, 1957), Part II; Bertrand Russell, *Principles of Mathematics* (Cambridge: Cambridge University Press, 1903), Part III. The discrepancy between the types of order described by Hempel and Nagel, and those given in the text, is due to the fact that I simplified the problem of quantitative methods by restricting my discussion to *unequal* quantities rather than covering all quantities, *both unequal and equal.*

these reasons the method of measurement should be subjected to careful, analytic study in order that the various kinds may be distinguished and the limitations of the method clearly indicated.

The English astronomer, A. S. Eddington, has a passage which indicates clearly the abstract nature of metrical concepts.

> Let us examine the kind of knowledge which is handled by exact science. If we search the examination papers in physics and natural philosophy for the more intelligible questions we may come across one beginning something like this: "An elephant slides down a grassy hillside. . . ." The experienced candidate knows that he need not pay much attention to this; it is the only put in to give an impression of realism. He reads on: "The mass of the elephant is two tons." Now we are getting down to business; the elephant fades out of the picture and a mass of two tons takes its place. What exactly is this two tons, the real subject matter of the problem? It refers to some property or condition which we vaguely describe as "ponderosity" occurring in a particular region of the external world. But we shall not get much further that way; the nature of the external world is inscrutable, and we shall only plunge into a quagmire of indescribables. Never mind what two tons *refers* to; what *is* it? How has it actually entered in so definite a way into our experience? Two tons *is* the reading of the pointer when the elephant was placed on a weighing-machine. Let us pass on. "The slope of the hill is 60°." Now the hillside fades out of the problem and an angle of 60° takes its place. What is 60°? There is no need to struggle with mystical conceptions of direction; 60° *is* the reading of a plumb-line against the divisions of a protractor. Similarly for the other data of the problem. The softly yielding turf on which the elephant slid is replaced by a coefficient of friction, which though perhaps not directly a pointer reading is of kindred nature. . . . And so we see that the poetry fades out of the problem, and by the time the serious application of exact science begins we are left with only pointer readings. . . . The whole subject-matter of exact science consists of pointer readings and similar indications.[8]

Eddington characterizes such knowledge as of form, not of content; of shadow, not of reality; of a skeleton, not of a flesh-and-blood creature. The abstract nature of measured values becomes even more clearly evident when we pass from direct measurements, such as that of length, to indirect measurements, such as those of density and hardness. And when we pass finally to measurements of intensive quantities, such as loyalty, love, piety, intelligence, and appreciation of beauty, we can see that the "descriptive"

[8] A. S. Eddington, *The Nature of the Physical World* (Cambridge: Cambridge University Press, 1928), pp. 251–252. Reprinted by permission of the publisher.

character of the process seems to have been almost completely lost. Whether we should therefore call measurement "descriptive" is largely a matter of how far into the area of indirect characterization we wish to extend the term. Here we are concerned merely with pointing out that measuring *is* a descriptive method, but capable of taking on an abstract character which makes the usage of the term "descriptive" seem somewhat out of place.

The Harvard physicist, P. W. Bridgman,[9] has stressed the fact that unless we are very cautious in our use of measuring techniques we may easily run into confusion with regard to what a measured value means. He points out that in one of the simplest of all measurements—that of length—many different processes are used in assigning values to the measured objects. If we measure city lots, lengths of moving street cars, lengths of fast-moving cathode particles, lengths of large tracts of land, solar and stellar distances, we use, in each case, a different process of measurement. Consequently, Bridgman asserts, we ought not to use the same word "length" in all these cases; we should invent special concepts such as "$length_1$," "$length_2$," "$length_3$," and so on, since a concept is synonymous with the set of operations by which it is defined. When we move to more complex measurements, such as those of density, conductivity, rate of growth, metabolic rate, time of food assimilation, degree of sociability or gregariousness, soundness of financial standing, degree of representation in governmental processes, and so on, we can see that the precision of information which is revealed by our measured value depends on how well understood the measuring techniques are, how dependable the instruments are, whether or not we are using the measurement process exactly suited to this particular object, and many other factors. Though Bridgman, being a physicist, is interested mainly in measuring methods, his prescription for defining concepts has been taken over by most of the sciences in one form or another and is commonly called "operationism." It applies to all of the descriptive techniques which we have discussed in this chapter, and is applicable in an even broader sense to explanatory methods as well.

Measurement is, of course, both physical and symbolic (or

---

[9] P. W. Bridgman, *The Logic of Modern Physics* (New York: The Macmillan Company, 1927), Chap. I. For a more extended discussion of *operationism* see A. Cornelius Benjamin, *Operationism* (Springfield, Ill.: Charles C. Thomas, 1955).

mental). In some cases it is dominantly the former—measuring lengths by yardsticks, measuring time by recording clicks of a pendulum, measuring elasticity of rubber balls by bouncing them, measuring amount of food consumption by weighing that which is given to the subject, measuring rate of the blood absorption by making successive blood tests, measuring an individual's poise by subjecting him to an interview, and measuring a student's knowledge of certain subject matter by giving him an examination. Frequently the measuring is done automatically by a properly constructed instrument; we determine the speed of automobile travel by merely looking at the speedometer, or the temperature by simply reading a number on a thermometer. In such cases it is permissible to speak of the numbers which appear as "signs" of the object measured, or as appearances which it manifests under the conditions when a measuring instrument is placed between the object and the observer. The relation between the object and its recorder is clearly physical in character, and this usage of "sign" would conform to that described in Chapter 3. But these physical numbers never enter into science unless they are "read" by an observer of some kind, and hence become sources of information about the object measured. One author[10] suggests that we use the word "number" as the actual plurality of objects, and the word "numeral" as the symbol for this plurality. Then the word "numeral" would be a verbal symbol for a plurality just as the word "white" would be a verbal symbol for the color of snow. This usage would avoid the difficulty of supposing that a measuring operation can be wholly mechanical or physical, without the presence of an observer who receives information from the number which he translates into a numeral. It would help also in taking care of metrical operations where, strictly speaking, no physical operation other than that of using the sense organs takes place. Such cases would occur when, for example, an experimental subject in psychology is asked to tell which of two tone pitches is higher, or whether one tone is twice as loud as another, or whether green is more like yellow than it is like red. Here metrical judgments, or, at least, quantitative judgments, are being made, yet no physical operations are involved.

Presumably there are three general types of process by which

[10] Campbell, *What is Science?* pp. 111–112.

numerals are attached to objects and events. One of these involves the establishment of a nominal scale, or a rank order.[11] This is closely related to our discussion of proper names on p. 50 and we can readily see that if the assignment of numerals to objects is more or less completely arbitrary, and the numerals assigned to the objects contain little or no information about the properties of the objects numbered, they can have little value in science. They serve only as directories or catalogues, and not as measurements.

On the other hand let us suppose that a group can be ordered by a non-quantifying ordering relation (one that is not directly interpretable as *greater than in respect to some quality*), and after the series has been established numerals are "attached" to the members of the series in the proper order. The numerals then become a manner of indicating position in the series, distance between elements, first elements and last elements, and they take on characterizing properties and constitute a type of measurement. Such ordered groups can best be explained in terms of examples.

Consider Mohs' standard scale of hardness, used by geologists. The class consists of the following substances, all of which are considered "hard" substances; the ordering relation is *is scratched by*, which, as can be readily seen, possesses the properties of asymmetry, transitivity, and connectedness; and this ordering relation is employed in such a way that *is scratched by* is represented by *is to the left of* on a horizontal line. The series is then indicated as follows: talc, gypsum, calcite, fluorspar, apatite, orthoclase, quartz, topaz, corundum, diamond. This, as it stands, is not a metrical series; in fact it is not even a quantitative series. But if we now interpret *is scratched by* as equivalent to *is softer than* it becomes a quantitative series. And if we now number the items from left to right as 1,2,3 . . . 10, then each numeral becomes a *measure* of the softness (or hardness) of the substance to which the numeral is attached. The limitations of such a series are obvious: it depends on the ten objects originally chosen to set up the series; it does not provide for any substances softer than talc; it does not provide a number for lead, which scratches talc but is scratched by gypsum; it measures degree of hardness merely by

[11] S. S. Stevens, "On the Theory of Scales and Measurements," *Philosophy of Science*, ed. by Danto and Morgenbesser, pp. 141–149.

"distance" separating the various members of the series; and so on.

Density of liquids could be measured by a similar process. If we take gasoline, alcohol, water, hydrochloric acid, and mercury, and set up a non-quantitative serial relation *floats in,* the above series will emerge. Now if we decide that *floats in* is equivalent to *is lighter than* (which is a quantifying relation), and then number our items 1,2,3,4,5 . . . we have a metrical scale and the number "1" would be a measure of the "lightness" of gasoline. (That this is not done in science is due to the availability of a more accurate measure of density in terms of mass divided by volume.) The Beaufort wind scale, devised by sailors before the invention of the anemometer, is as follows: calm (0); light breezes (1,2,3); moderate winds (4,5); strong winds (6,7); gales (8,9); storms (10,11); and hurricanes (12). The ordering relation is here *less strong than.*

Such scales, when there is available a quantitative correlate to the non-quantitative ordering relation, are frequently called "intensive quantitative scales." Whether such series are metrical depends upon how "reasonably" one can attach numerals to objects in the quantitative series. It would be interesting in this connection to consider the examples of series given by Larrabee (see p. 71) and ask whether they could be metricized. The arrangement of students in our schools into a series on the basis of the ordering relation *more intelligent than* is actually done with, perhaps, only moderate success, by translating this relation into *answers more of a given set of examination questions than,* and then attaching either the numbers 100,99,98 . . . 0, or the letters A,B,C,D,F, to the students on the basis of this serial arrangement. As was suggested, metrical series of moral, aesthetic, and religious values would seem to be so artificial as to be practically useless.

Much more important in science are those series characterized as "extensive quantitative scales." The best approach to a consideration of such series is to take the following example: Let us suppose that we have a class of objects of different weights (masses), and set up the ordering relation *outweighs.* This relation can be applied by saying that x outweighs y if, when the objects are placed into opposite pans of a balance in a vacuum, x sinks and y rises. On the basis of this operation we can establish the quantitative series whose ordering relation is *weighs more than.* But the peculiar character of this series is that x *weighs more than* y

is equivalent to *the weight of x is equal to the weight of y plus p.* In other words, extensive quantities are such that any greater quantity can always be obtained from any lesser quantity by an additive operation. The additive operations must possess the same properties as those exhibited by the mathematical "+" sign. Some of the most common extensive quantities, in addition to mass, are lengths, areas, volumes, durations, electrical resistances, and (in certain instances) velocities. In the cases of masses, adding x to y means *putting on the same pan of a balance*; in the case of lengths, addition means *putting end-to-end in a certain specified way*; durations can also be added by starting one duration at the same time that another duration ends (this is not the case when a judge punishes a criminal who has committed two crimes by giving him a sentence of three years for one offence and five years for the other, the two sentences to run concurrently); velocities can sometimes be simply added, as in the case of the determination of the velocity of a train conductor relative to the earth by taking the mathematical sum of the forward velocity of the train and the forward velocity of the conductor on the train relative to the train: in certain other cases velocities cannot be added.

We have one final distinction which must be clarified in the discussion of measurement. A process of measurement is called "fundamental" if the assigning of a number to a quality or object is direct and not based upon any previous assignments of numbers to other qualities or objects. Measurement of the latter kind is called "derived." The best examples of the former are the measurement of length, duration, and mass. (The problem of the relation between extensive quantities and quantities which are fundamentally measurable is a complicated one, and cannot be discussed here.) While the measurements of length, duration, and mass appear to require previous measurements, since they employ such established standards as centimeters, seconds, and grams, nevertheless we can readily see that the attaching of numerals to the standards is not itself an act of measurement but an arbitrary decision. We do not determine that a standard length is one centimeter by measuring it; we stipulate arbitrarily that a given length is to be taken as a standard. Thus the assignment of the symbol "20 centimeters" to the length of an object depends upon no previous measurement, and is consequently a fundamental measurement in the sense in which the term is here used.

Derived measurement takes place when numbers are assigned not directly to the quantities but to other quantities having a specified relation to the quantities which are to be measured derivatively. There are two ways (not sharply distinguishable) in which this can be done. The first way is *by definition*. "Density" is defined as mass divided by volume; "force" is sometimes defined as the product of mass and acceleration; "velocity" is defined as space divided by time; and "I.Q." is defined as mental age divided by chronological age multiplied by 100. The second way of obtaining derived measurements is *by law* or *correlation*. "Temperature" is measured by the height of a column of mercury in the ordinary thermometer; "elevation above sea level" is measured by air pressure in the barometer; angular spread is measured in trigonometry by, say, the tangent of the angle. In the first cases the correlation is based, respectively, upon physical laws concerning the effect of heat on material bodies and the relation between altitude and atmospheric pressure; in the third case a trigonometric theorem is involved. Frequently a measurement derived by law becomes transformed into a measurement derived by definition. Many loose correlations could be employed to set up very rough methods for obtaining derived measurements. The beauty of paintings could be measured in terms of price paid for them; pleasures could be measured in terms of the money required to produce them or in terms of some complex function of their intensity, duration, certainty, and purity; happiness might be measured by the quotient of possessions over wants or needs; the respect which we have for an individual is roughly the quotient of his successes over his boastfulness. None of these would even approach the precision which is demanded by science, but they provide opportunities for seeing how derived measurements are obtained.

The result of the activities of observing, interpreting reports, classifying, associating, ordering, and measuring is a collection of symbols which, because of the operations involved in their derivation, may be said to be broadly descriptive of the objects and events from which they have been obtained. For a convenience in terminology we shall henceforth call such a body of symbols in their relation to the world a "descriptive science." As we shall see later, it should not properly be called a finished science since it usually represents either a stage in the development of science or the "factual portion" of a more complex science which contains theories and hypotheses. The symbols constituting a descriptive science are, as we have seen, mainly proper names, icons, and characterizing symbols. But whatever they may be, the important feature of the system as a whole is the "directness" with which it attempts to portray the obvious facts of our experience. Such a science, of course, contains no symbols for electrons, the inside of the earth, the other side of the moon, perfect levers, or the ether—in fact no symbols for any of the entities of the world which are unobservable because they are too small, or too remote to be detected by our present instruments, or because they are hidden from us by opaque objects, or because they are not real entities

FIGURE 7
DESCRIPTIVE SCIENCE

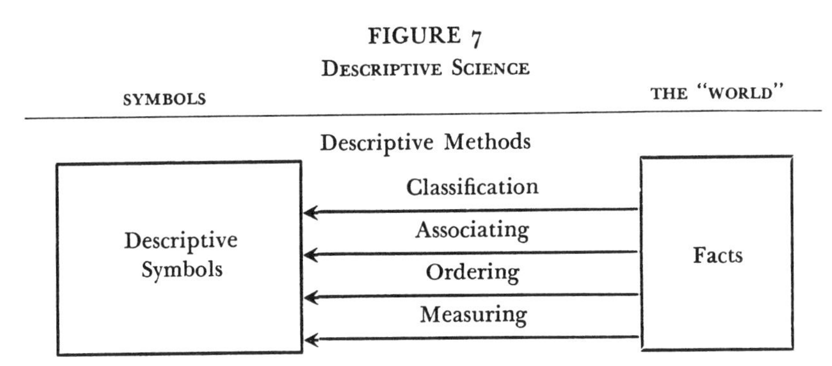

at all but fictions and idealizations. Furthermore, when emphasis is laid on the directness with which such a scheme represents the world, we should not lose sight of the fact that even these descriptive processes are not so direct as we might wish them to be. Observation, as we have already seen, is subject to many types of error; reports are reliable only when we have correctly interpreted what has been reported, that is, when we have made due allowance for the errors and inaccuracies of the recording instrument, human or mechanical. Classification and correlation involve the comparison of objects, the isolation of similar properties, and the invention of general names or concepts which are presumed to represent these characteristics. Ordering and measuring are full of pitfalls; the complicated mental processes involved in these activities leave much room for error. When we say, therefore, that a descriptive science represents the realm of observed facts *directly*, we mean only that it is the most direct representation possible in view of the nature of the knowing process. We can find no manner of getting "closer" to facts than to observe, manipulate, and gather reports about them, and to classify, associate, order, and measure them. The nature of a descriptive science is represented in Figure 7.

5

The Mystery of Scientific Discovery

Before passing to a consideration of the role of theories in science we shall do well to see where we stand. We have attempted to explain what is meant by the method of confirmed hypotheses as the peculiarly scientific method. This led to a realization that science must rest ultimately on facts, both as a point of origin and as a point of final reference. The techniques for getting facts proved to be mainly observation and the interpretation of reports, manipulatory and experimental methods, and finally descriptive procedures, which included classifying, associating, ordering, and measuring. The result of these activities is the establishment of a symbolic scheme which claims to portray the characteristics and relations of the facts which have been discovered. It can be considered a good scheme to the extent to which it describes all the facts, and therefore is at least as extensive as the realm of facts, and only the facts, and therefore is no more extensive than this realm. Such a science represents the result of the application of the most basic stage in the method of confirmed hypotheses.

1. Is Science Descriptive?

Now we cannot pass to a consideration of the further development of this method without recognizing and replying to a criticism which will immediately arise in the minds of many people. Just why, they will ask, is the stage of description considered to be the *first* stage in the development of science and not the *last*? Isn't the ultimate goal of science the establishment of a symbolic scheme which describes the total world of facts? When we have observed all the objects and events in nature and society, when we have achieved a complete and final description of them, including a precise formulation of their manner of behaving under varying circumstances, when we have shown how they are correlated with one another, is there anything left for science to do? The job of science is to tell us what everything is and how it acts; a completed description does just this. This seems to have been the view of science held by Bacon. He argued, as Jevons points out, that science was a "kind of bookkeeping. Facts were to be indiscriminately gathered from every source, and posted in a ledger, from which would emerge in time a balance of truth."[1] What would be the advantage, therefore, in going on to theories and hypotheses, which involve positing mysterious powers and entities beneath and behind nature—entities whose nature we can never know and whose existence we can never prove? Why clutter up our knowledge with conjectural notions when we already have a science with the required certainty?

Furthermore, argue these critics, to consider description as the first stage in science is to commit a fundamental error. Primitive science did not begin with description at all. Early science had very few facts and was made up almost wholly of imaginative and fanciful notions which posited the existence of powers and beings who inhabited the natural realm and produced the sun and the storms, life and growth, disease and death, good and bad luck in hunting, and all the other events vital to the life of primitive man. Man projected himself into nature and explained everything in animistic or anthropomorphic terms—gods, demons, fairies, and other spirits.[2] Most of the facts he was trying to ex-

[1] W. S. Jevons, *Principles of Science* (London: Macmillan & Co., Ltd., 1907), p. 576.

[2] *The Positive Philosophy of Auguste Comte*, ed. Harriet Martineau (London: G. Gell and Sons, 1896), pp. 25–26. In this passage Comte describes his famous *Law of the Three Stages*. According to this law, man, in his intellectual development from

plain were not real facts but pseudo facts, and most of the explanatory entities which he employed for the purpose of accounting for these apparent facts were such as cannot survive scientific scrutiny today. The earliest stages of science, therefore, were anything but scientific.

Not only is it a mistake to suppose that science began with description, the criticism continues, it is equally erroneous to believe that science aims at explanation.[3] True, the early scientist soon realized the futility of trying to explain nature by gods and demons, for this enabled him neither to understand it nor to control its behavior. Consequently he readily abandoned these primitive notions when he began to notice certain recurring events, such as the rising of the sun and the sequences of the seasons; when he began to acquire power over nature and could build his own fires, cure many of his diseases, and cultivate his own crops; and when he found that propitiation and sacrifice failed to produce the desired results. But the realization of the futility of such explanation surely carried with it a realization of the futility of *all* explanation. Primitive man made his mistake, the argument goes on to show, not *in the way* in which he tried to explain nature, but in the very attempt to explain nature *in any way*. The question *Why?*, he argued, is really unanswerable, and the question *How?* should obviously be put in its place. The former leads merely to endless search and produces no concrete results. The latter, on the other hand, is definitely promising since it asks simply for a description of causes and conditions under which natural events occur. The argument concludes, then, that scientific knowledge can attain maturity only if it deliberately abandons the attempt to explain nature and substitutes for this the endeavor to obtain the most adequate and most complete description of how natural processes take place.

---

a primitive stage to modern science, attempted to explain nature first in theological and animistic terms; then, recognizing his lack of success, abandoned these projections of himself into nature and replaced them by depersonalized forces, abstract powers, and potentialities; and finally, realizing the futility of both methods, gave up entirely the attempt to *explain* the world and decided to remain satisfied with a complete and accurate *description* of the way in which it manifests itself to our senses, when these are subject to careful direction and control.

[3] This word will not be subjected to careful examination until we reach Chapter 6. Meanwhile we shall use it in a very crude sense, as employed by the common man, and as roughly equivalent to "accounting for," "making understandable," or "reducing to something better known."

Finally, the argument concludes, even granting that science *does* try to explain nature as well as to describe it, no sober-minded scientist would ever begin his investigations by going out into the world, gathering facts at random, inspecting them with various recording devices, and performing haphazard experiments on them—all with a view to classifying them, showing what correlations might be found to hold among the properties of the many objects collected, ordering them, and measuring them. Nor would he then, having done all this, sit down and patiently wait for some hypothesis to occur to him which would make sense out of what he had been doing. Science does not gather facts in order to theorize; for how could it, apart from theories, know what facts to collect and what to disregard? Facts are noticed only because they support or contradict some projected theory. Description cannot be the first stage of science for we shall never know what is important enough to be worth describing until we have formulated some problem to which it is relevant, positively or negatively. It is absurd, therefore, to speak of description as being the *first* stage of science.

2. Positivism and Its Criticism

The force of this criticism cannot be denied. In fact, many of its arguments are sound, though there is no necessary common point of view among those who accept them. In order to characterize those who maintain this conception of science we shall call them the "traditional positivisits"—"traditional," to distingush them from the more recent "logical positivists," many of whom would not accept the arguments which have been here presented. The traditional group[4] includes Auguste Comte, Karl Pearson, Ernst Mach, and some more recent figures who, without calling themselves positivists, tend to relegate all hypotheses to an inferior

[4] The article on "Positivism" in the 1947 edition of the *Encyclopedia Britannica* claims that "physical scientists generally view the universe from the positivistic standpoint." I am sure that this statement was not true at the time it was published, nor is it true today. I am happy to state that the editors have corrected it in recent editions. Among the traditional positivists are the following: Auguste Comte, *op. cit.*; Karl Pearson, *The Grammar of Science* (3rd ed.; Cleveland: Meridian Books, Inc., 1911); and Ernst Mach, *The Analysis of Sensations* (Chicago: Open Court Publishing Co., 1914). See also *Theories of Scientific Method*, ed. R. M. Blake *et al.* (Seattle: University of Washington Press, 1960), especially Chap. 12, "Pearson's Search for a Method."

position in science by calling them "conceptual shorthand," "linguistic tools," "heuristic devices," "pragmatic fictions," and to look upon them merely as convenient instruments for grouping and ordering facts rather than as real but unobservable objects hidden in the recesses of nature. The members of the group are afflicted with what Bavink calls "hypotheseophobia"[5]—fear of hypotheses—and hold in common the view that the *best* science is descriptive, and when it departs from this ideal it tends to deteriorate. Many scientists who accept traditional positivism prefer today to call themselves "pragmatists," since they reject completely the extreme realism of Bavink, for example, who says that "*atoms are just as real things as cannon balls or grains of sand, as waves on water or mountains* [Bavink's italics]."[6] They wish to play more cautiously with science, and to refrain from projecting into nature any explanatory device, however successful it may be in its role of helping us to grasp the problems and puzzles of the world. Consequently they employ such concepts merely as linguistic tools, enabling us to make predictions as to further facts which we hope to discover, but not as constituting the actual furniture of the world.

The traditional positivists are to be sharply distinguished from the more recent logical positivists (Carnap, Feigl, Schlick, Philipp Frank, Otto Neurath, and many others). In fact the latter group has tended to abandon the self-characterizing term "logical positivism" for the terms "logical empiricism" or "scientific empiricism" simply because of the tendency on the part of the public to confuse them with the traditional positivists. They tend to take a neutral position with regard to the status of hypotheses, arguing that the question as to whether they exist or do not exist is a metaphysical pseudo problem, whose solution is in no way required for the progress of science. Hence they confine their attention largely to problems of the *logic* of science, and even, somewhat more narrowly, to the logic of the *language* of science.

Now that we have located properly the advocates of the strictly descriptive view of science, we shall turn our attention to replying to their arguments and attempting to evaluate the point of view

[5] Bernhard Bavink, *The Natural Sciences* (New York: The Century Co., 1932), p. 39. Bavink evidently took the term over from Ed. von Hartmann, though he does not give the citation.

[6] *Ibid.*, p. 29.

which they accept. Let us consider, first, the idea that descriptive science is the culmination of man's long search for a solution to the problem of the nature of the universe and man. Certainly there can be no doubt about the fact that science as we know it today was preceded by a prolonged period in which inaccurate observation, wishful thinking, and uncontrolled speculation were the accepted methods of knowing. Primitive science was not at all like modern science. Not only were the instrumental devices and measurement aids lacking—tools which are essential to any well-equipped laboratory today and which contribute so much to objectivity and to the elimination of the personal equation—but what we have called the "scientific spirit" was completely lacking. Early investigations into nature were neither objective, unemotional, rigorous, nor controlled. All this must be granted as historically accurate as far as we are able to draw any conclusions concerning the attempts of early science to understand nature.

But why should such attempts be called "science" at all? Wouldn't it be well to reserve the word for that pursuit of knowledge which arose *after* man had developed a conscious recognition of logical techniques and had made a more or less deliberate attempt to control his speculations by the somewhat rigid rules of reasoning? To be sure, the dispute is simply one of words, and to this extent is not critical. But certainly to call something "science" when it lacks the minimum requirements which we associate with that study is confusing. We should recognize that what science began with was not science at all.

Furthermore, in arguing that description is the first stage in the development of science we must insist that we are using the word "first" in the logical and not in the temporal sense; the issue which we are discussing is one of method, not one of history. Certainly it would be absurd to suppose that scientists spent, say, the first thousand years of the Christian era in the gathering and correlating of facts, and then, by common consent, shifted their activities from the accumulation of data to the devising of theories and other explanatory notions. Science does not develop logically but, like Topsy, just grows. Description is certainly an early stage in science in the sense that all theorizing depends on facts. *With* facts, the speculative activities can go ahead; *without* them, they cannot. The accumulation and accurate description of facts, therefore, is logically presupposed in all attempts to explain. *Some* facts must be given in advance of all theorizing; otherwise there would be nothing about which to theorize. And certainly anything that can be done to clarify the facts, and gather more information about them will serve to speed up the suggestion of theories.

Finally, while there was, historically, an abandonment of the primitive modes of explanation in favor of the more critical, logical methods, this usually involved merely a change in the *kind* of explanation, and not a complete rejection of the *fact* of explanation. As man gained intellectual maturity he saw with greater and greater clearness that attempts to explain the world in terms of feelings, wishes, hopes, and fears were doomed to failure. The world proved to be quite unconcerned with man's desires, and he soon learned that he could neither understand it nor gain control over it so long as he persisted in projecting himself into nature and attempting to interpret it in terms of his own personality. But this did not lead him, in most cases, to give up completely the attempt to explain the world and his place in it. He merely sought new kinds of explanations—explanations in terms of abstract depersonalized forces and powers, like gravitation, chemical affinity, electrical attraction, heredity and growth, and stimulus and response. He did not abandon the attempt to answer the question, *Why*? He merely sought a new type of answer to the old and ever recurring question.

The traditional positivists, therefore, seem to be wrong in insisting that description represents full-grown science. Facts must be collected, of course, before there can be science, and they must be classified and correlated. But a descriptive science is not a mature science; rather it is science in its youth—science still exhibiting the characteristics of childhood but showing also the promise of adolescence. When an inquiry has become descriptive it has taken on only the minimum features of a genuine science. It gives us something to talk about—something which arouses our curiosity, like symptoms told to a doctor. But not until the facts have been accurately observed, reported, perhaps manipulated in various ways, and organized, are we ready for the real task of science—that leap into the unknown which provides us with an explanation in terms of hypothetical and theoretical notions.

3. The Urge to Explain

If we look at the problem somewhat more analytically we can see, perhaps, why science cannot rest satisfied with pure description. In the first place, no actual descriptive science *has* all the facts. While we might admit, conceivably, that a completed descriptive science would do all that is required of a perfect science, the plain fact of the matter is that no existent science, either descriptive or explanatory, even approximates to this ideal. The discrepancy between the facts which we know and those which we shall know in the future, or conceivably might know, is appalling. Furthermore, as we have already intimated, frequently the most effective method for increasing the range of facts is to abandon description temporarily and anticipate other facts through the method of hypothesis. To be sure, we run certain risks this way: we often get off on the wrong track and anticipate facts which prove not to exist at all. But, on the other hand, we do make many fortunate guesses and succeed not only in finding facts which might have completely escaped our notice for long periods of time but also in discovering facts which we should never have found at all since they had to be created under the guidance of theory. The history of science abounds in illustrations of facts which were discovered simply because a previously suggested theory told the scientist where to look and what to look for. The discovery of the planet Neptune[7] might have been delayed for years had not Adams and Leverrier concluded that it must exist in order to explain the perturbations in the motion of the planet Uranus; and the creation of the atomic bomb would have been impossible without the theoretical work of Einstein and other physicists.

In the second place, one may well talk about abandoning the attempt to explain, but man will not do this. He is incurably speculative and no admonition to confine his knowledge to observable facts will be effective. In spite of having committed many errors in the history of thought he has not learned to (perhaps he does not want to) control his imagination by the techniques of logic, and give up random speculation and wishful thinking. Certainly some advance over his more primitive ancestors is evident in this matter. But to ask man to give up completely all attempts to explain; to forbid him to ask for reasons; to stifle his curiosity

[7] W. T. Sedgwick and H. W. Tyler, *A Short History of Science* (New York: The Macmillan Company, 1939), pp. 390–391.

by edict—these are doomed to failure simply because man is what he is. Any parent knows how persistently his children at a certain age ask the question, *Why?*[8] Add to this the fact that man does get satisfaction from his attempts to explain even though he recognizes that many such efforts end in failure and that no final explanation will ever be achieved. Much of the pleasure derived from reading a detective story lies in the identification of the criminal before the author discloses him to us; even if we fail we derive satisfaction from the effort. In science, where our attempts to explain the ultimate nature of matter have led from molecules to atoms, and from atoms to electrons, protons, neutrons, positrons, and still others which only the future can reveal, the enterprise is more exciting just because it seems to be endless. No one can legislate such speculations out of existence.

In the third place, descriptive sciences, when they are examined from a certain point of view, often exhibit a very remarkable feature which could not have been anticipated from our knowledge of the way in which these sciences have been derived. It shows that they are, in a sense, vague anticipations of explanatory sciences, and that the germs of their later development into full-grown theoretical sciences may already be present. Let us illustrate this in terms of a very simple descriptive science. A sociologist submits to a group of married men a questionnaire in which he asks them to provide certain information about themselves—whether they have had a college education, whether they are now earning over $8,000 a year, whether they hold executive positions, how many children they have, etc. Let us suppose that when the results from nine questionnaires have been tabulated the following relevant information appears:

INDIVIDUALS REPORTING	COLLEGE EDUCATION?	EARNING OVER $8,000 PER YEAR?	HOLDING EXECUTIVE JOBS?	NUMBER OF CHILDREN?
Smith	Yes	Yes	Yes	4
Brown	No	Yes	Yes	2
Jones	Yes	Yes	Yes	8
Williams	No	No	No	3
Burch	Yes	Yes	Yes	5
Green	No	No	Yes	6
Walters	No	Yes	Yes	4
Oliver	No	No	Yes	1
Mathews	Yes	Yes	Yes	7

[8] A. D. Ritchie, *Scientific Method* (London: Kegan Paul, 1923), remarks that the Greeks were always children, presumably because they persisted in asking the ques-

From these data we can draw the following conclusions:

A. All who have had a college education are earning over $8,000 a year.

B. All who have had a college education are holding executive positions.

C. All who are earning over $8,000 a year are holding executive positions.

These statements are all descriptive correlations in the sense that each of them indicates that all of the members of the group which have a certain property or characteristic have also another attribute. They are all true in the sense that they have been obtained by direct study of the answers given on the questionnaire. While each of the statements asserts a correlation between properties, we do not at present know that the statements themselves bear any significant relation to one another over and above the fact that they are all true.

But if we re-arrange the statements as follows we make an important discovery:

A. All who have had a college education are earning over $8,000 a year.

C. All who are earning over $8,000 a year are holding executive positions.

B. All who have had a college education are holding executive positions.

A certain logical structure, called a "syllogism," emerges. For we can now see that *if* A and C are true *then* B must be true. If we had known in advance that A and C were true, but had not known that B was true we could have inferred its truth deductively. All such structures involve statements which are called "premises" and statements which are called "conclusions," and the premises are said to "imply" the conclusions. The inference is called "deductive" in order to distinguish it from "inductive" inference, a form which we shall examine later.

4. Detection of Deductive Structures

The significance of this discovery can be clearly seen if we ask how it changes our understanding of descriptive science. It sug-

tion, Why? But he cannot resist adding, "This question, as every adult knows, is both wicked and preposterous" (p. 2).

gests at least the possibility that *all* statements in such a science might be interrelated according to the same logical pattern, i.e., that the science might be divided into two groups, premises and conclusions, such that certain members of the first group, when combined, would imply one or more members of the second group, and that each member of the second group would be implied by a combination of members of the first group. The former could then be called "implying descriptions" and the latter "implied descriptions." All of the statements are *descriptively* derived since they are obtained by direct examination of the data; but the implied descriptions are also *logically* derived because they can be deductively inferred from some of the implying descriptions. The important point is that the implied descriptions *need not* have been descriptively derived, since they could have been deductively derived from the descriptively true premises. Thus we could have obtained them by "pure thinking" rather than by a further exploration of the world.

As a result of this discovery we find ourselves provided with a very important technique by which we may expand our descriptive science without bothering to look for further facts in nature; we have a method by which we may anticipate what the world will disclose to us if we are interested in obtaining more data. Let us return to the questionnaire of our sociologist. Two other descriptive truths which we can obtain from the data are the following:

D. No man has more than eight children

E. No man is childless

From these statements we can infer deductively that there are at least two men who have the same number of children. For if we attach a number to each man (there are nine in all), indicating the number of children he has, Smith will be #4 since he has four children, Brown will be #2 since he has two children, and it can be readily seen that since there are no childless men and the number of *men* is greater than the *number of children which any man has,* there must be two men with exactly the same number of children. In the case of the simple questionnaire which we have examined this information could have been readily obtained by direct inspection of the answers given by those reporting, and if we had proceeded in this way our descriptive science would have grown by the usual method. But a questionnaire containing twenty-five items might have caused us some difficulty. However, if we knew that no man had more than twenty-four children (a plausible assumption on other grounds!) and that every man had at least one child we could have drawn the same conclusion. Thus by means of logic we are able to get advance information concerning the nature of the world; we can know of the existence of the facts before we have discovered them.

Once we have appreciated the fertility of this method for expanding descriptive science we are in a position to ask whether there may not be a further technique, based upon the same logical organization, which would permit us to "work in the other direction." We have seen that a descriptive science may consist of true statements falling into the two classes of implying and implied statements such that members of the second class may be logically deduced from members of the first class. We have just learned that certain statements in a descriptive science, not known to fall into either of these classes, may, upon inspection, prove to imply certain other statements. The implying statements thus permit us to deduce new truths, and these can be added to our descriptive science. This entitles us to extend it and increase its scope: from implying truths we can derive implied truths. Is it now possible that other statements *in* our descriptive science, not known to be either implying or implied statements, might suggest to us some heretofore unknown statements which imply *them*? If so, our descriptive science could be similarly extended to include these newly discovered statements: from implied truths we could "derive" implying truths.

But apparently this is not possible. When P implies Q we can say that if P is true Q must be true, but we cannot say that if Q is true P must be true. Or, thinking of the "movement of our thought" from one statement to the other, to go from P to Q is not the same as to go from Q to P. There are, of course, cases in which P implies Q *and* Q implies P. In these instances the "direction" of our thought is the same in both cases. But then the distinction between implying statements and implied statements disappears, for each statement is of both kinds. Since we wish to preserve the distinction between the two methods of extending descriptive science we shall simply adopt the convention that when we assert that P implies Q we shall mean that Q does not, so far as we know, also imply P, i.e., that P is uniquely an implying statement and Q is uniquely an implied statement. The problem, then, is to determine whether there is any method by which a given statement in a descriptive science may be "made into" an implied statement, i.e., may be discovered to be a Q-statement such that there is a P-statement which implies it.

The simplest answer to this question is that we do repeatedly go from Q-statements to P-statements. In fact, whenever we have a statement whose truth we want to establish, either to inform ourselves or to inform others, we search for premises from which it can be deduced. For example, if I am engaged in an argument and trying to prove to my opponent that a given statement, "This piece of gold is malleable" is true, I search for another statement, "All gold is malleable and this is a piece of gold," which he will accept and which implies my statement. My original statement then becomes a Q-statement and the discovered statement becomes a P-statement. If I can convince my opponent that P is true, and that P implies Q, he must, if he is a rational creature, admit also that Q is true.

Among the recent writers in the philosophy of science, the one who most clearly develops the point of view I am here presenting is Norwood R. Hanson.[9] In fact he dignifies the movement of thought from data to hypothesis not only by calling it an "inference" but by specifying its peculiar properties and calling it "retroductive inference." Many writers would hesitate to list it among the inferential processes because this seems to place it in the field of logic whereas it is purely and simply a matter of psychology. This would be a serious objection were the fields of logic and psychology as sharply distinguished as many logicians mistakenly suppose them to be. I am inclined therefore to agree with Hanson in giving it a special name which will indicate the essential role which it plays in all science. C. S. Peirce[10] calls it "abduction," which he distinguishes from "induction." I shall call it "induction," using this word in the broader of its two generally accepted meanings. This is the meaning which we com-

[9] Norwood R. Hanson, *Patterns of Discovery* (Cambridge: Cambridge University Press, 1958); Norwood R. Hanson, "Is There a Logic of Scientific Discovery?" *Current Issues in the Philosophy of Science*, ed. H. Feigl and G. Maxwell (New York: Holt, Rinehart & Winston, Inc., 1961), pp. 20–42; Norwood R. Hanson, "Retroductive Inference," *Philosophy of Science: The Delaware Seminar*, ed. Bernard Baumrin (New York: Interscience Publishers, Inc., 1963), pp. 23–27.

[10] C. S. Peirce, *Collected Papers*, ed. C. Hartshorne and P. Weiss (Cambridge: The Belknap Press of Harvard University Press, 1932), V, 413–432.

monly have in mind when we say that the sciences employ the inductive-deductive method; my choice of the phrase "method of confirmed hypotheses" as a general designation for the method of the sciences is based on the fact that from data we *induce* hypotheses, and from hypotheses we *deduce* predictions by means of which we hope to confirm our hypotheses. The narrower conception of induction as "reasoning from particular cases to general laws" falls under the wider meaning, by being a special case. When our original data are particular correlations and what we are seeking is a general law, then our law becomes the hypothesis which we induce from the special cases. But there are many scientific hypotheses which are not laws, yet we are quite entitled to say that we induce them from the data; for when we have found them they do imply the descriptive truths characterizing the data. To summarize: given Q, if we wish to find a P which implies it we "move" *inductively*; and given P, if we wish to find a Q which it implies we "move" *deductively*.

What must be clearly recognized are the important differences between the two movements. (1) When we go from P to Q we know that if P is true Q must be true; but if we are given that Q is true we cannot necessarily infer that P is true—it is only possibly or probably true. If Q is a statement in a descriptive science and we are able to find a P which implies it, we can say only that we have provided ourselves with a plausible guess and our efforts must now be turned to a further examination of P for the purposes of removing its uncertainty and rendering it sufficiently probable to include it among the affirmations of our science. (2) There are logical rules for going from P to Q, e.g., the rules of the syllogism, but there are no rules for going from Q to P; the movement in this direction rests on the skill and ingenuity of the investigator. For this reason I shall often use the word "insight" in the following pages, as a synonym for "induction." This will stress the absence of rules and the necessity for admitting the spontaneous and uncontrolled character of the movement.

This inductive act is well illustrated by the manner in which the detective in the mystery thriller endeavors to find the culprit. He proceeds by a careful examination of the condition of the corpse, the time and place of the crime, and other relevant information to determine who committed the murder. The data in the problem constitute the evidence from which, by an act of induction or insight, the detective passes to the hypothesis stating who is responsible for the act. Authors of mystery stories frequently call this "reconstructing the crime" — a fortunate expression, since it suggests that the whole thing is a matter of artistry and genius rather than of pure logic. Many authors, among them A. Conan Doyle, call this movement of thought "deduction." This is confusing. While there may be, as we have just seen, implications from facts, generally the evidence does not imply the hypothesis; on the contrary, the hypothesis implies the evidence. We arrive at our hypothesis by a "leap of thought," knowing that if we have made the correct guess our data can be properly explained.

For all these reasons the claim of the traditional positivist that descriptive science represents the final rather than the initial stage of science seems untenable. Descriptive science, in fact, is very unstable. Even as descriptive science it tends to grow and increase in stature; it tries to encompass more and more facts, and to achieve greater and greater accuracy in the portrayal of the facts it has already at hand. Furthermore, even as descriptive science it exhibits the rudiments of logical structure among its elements, and it strives to improve this logical organization by introducing new elements which will increase its coherence. It aims, for example, to eliminate what are sometimes called "brute facts" — facts which seem undeniable but which appear to have no relation to the rest of the facts dealt with by the science. If we grant that through the speculative activity of the mind we can further these basic demands of descriptive science, and if we agree also that the speculative activity of the mind is so essentially human that we cannot possibly prevent it from functioning, there seems no longer any important reason for accepting the claim of the traditional positivists. Description *is* an important stage in the development of science, but must be supplemented by explanation; only in explanation is science really coming into its own.

5. The Unsolved Problem of Insight

We are not yet ready to consider in detail the precise manner in which a descriptive science becomes transformed into an explanatory science. This will be done in our next chapter. Here we wish only to point out how little we really know about insight. From the psychological point of view it is simply the speculative activity of the mind endeavoring to find reasons for facts which do not wear their explanations on their coat sleeves. Its motive is simply curiosity and the unconquerable desire to understand; its reward is the satisfaction achieved when brute facts are brought together under a unifying conception, when mysterious happenings yield to rational interpretation, and when predictions which are the necessary consequences of the imaginative conception take place as anticipated.

One could wish that a method which has so much to contribute to the advancement of science would yield to adequate analysis. One could even hope that the procedure would be capable of formulation according to rules which would so direct the activity of the scientist that he could "grind out" the required hypothesis whenever the occasion demanded. The truth is, however, quite otherwise. We know neither what takes place in this movement of discovery, nor how to control it and foster it. We recognize our failure by calling it a "mystery."[11] It is as puzzling and unexplainable as artistic inspiration, as unpredictable as intuition, as incapable of analysis as genius. All that we know about it can be summed up in a few statements: It is essential to science. It does occur on numerous occasions. It is distributed very unequally among men. It produces most significant results when preceded by a long and persistent study of facts.

In an often-quoted passage Graham Wallas[12] analyzes the act of discovery into four stages which he calls "preparation," "incubation," "illumination," and "verification." The last of these four terms does not properly belong in our discussion of this chapter, since we are here concerned not with how we confirm hypotheses but with how we get them in the first place. The introduction of the term "preparing the imagination" is probably due to Tyndall.[13] Before discussing the incubative and illuminative processes,

[11] The attempts to solve this mystery have been many and various. I shall list only a few of the recent contributions to the problem. René Taton, *Reason and Chance in Scientific Discovery* (New York: Philosophical Library, Inc., 1957); A. L. Porterfield, *Creative Factors in Scientific Research* (Durham: Duke University Press, 1941); I. B. Cohen, *Science, Servant of Man* (Boston: Little, Brown & Co., 1948), especially Part I, Chap. 3; W. I. B. Beveridge, *The Art of Scientific Investigation* (New York: Random House, Inc., 1957), Chaps. III, V, and VI; and Mario Bunge, *Intuition and Science* (New York: Spectrum Books, Prentice-Hall, Inc., 1962).

[12] Graham Wallas, *Art of Thought* (New York: Harcourt, Brace & Co., 1926), Chap. IV.

[13] John Tyndall, "The Scientific Use of the Imagination," *Fragments of Science* (New York: Appleton, 1898), II, Chap. viii.

we shall ask what can be meant by preparing the imagination. Broadly speaking, all processes of gathering the facts, and all descriptive operations serve precisely this role. Unless we get the facts through observation and reports we have nothing for insight to work upon. (A famous cookbook prefaces the recipe for rabbit stew by the statement, "First, catch the rabbit.") By experimental manipulation of these facts we increase our familiarity with them. By describing them we arrange and order them in such a way that they begin to take on the character of a system. In all these attempts we are simply recognizing that insight never arises out of either ignorance or confusion; men of genius are both well informed as to facts, and orderly in their thinking. Although we cannot guarantee that when we have a mature descriptive science there will occur inevitably that flash of insight which transforms it into an explanatory science, we can rest assured that *unless* we have met this minimal condition genius cannot operate.

One of the most striking aspects of the scientific method is the frequency with which flashes of insight occur in situations which are purely and simply accidental—situations in which there seems to have been no "preparation" whatsoever. These are so common in science that writers on scientific method have had to invent a term to describe them: "serendipity."[14] Serendipity is accidental discovery, i.e., the gaining of a new hypothesis from an experiment which was set up for a totally different purpose, or from the purely chance coming together of objects or events which produced effects and disclosed properties never before known. While the term has never been precisely defined, it is commonly used in both a broad sense and a narrow one. Broadly speaking "serendipity" refers primarily to the suddenness or unexpectedness (at the time) of the act of discovery; that discoveries do occur to the scientist outside his laboratory and in moments of recreation is a noteworthy fact, and we shall turn to illustrations of this usage of the word immediately. According to its more restricted meaning the word refers to discoveries totally unrelated to *any* problem which the scientist is working on at the

[14] The word has its presumed source in a story, "The Three Princes of Serendip," which was read at one time by Horace Walpole. Because the three princes were always "finding things they were not in quest of" Walpole invented the word to characterize this type of unexpected discovery. For a further discussion of the origin of this word see *Science*, Vol. 142, No. 3593, p. 621, and Vol. 143, No. 3603, p. 196.

moment, and hence usually of great importance in the forward movement of science.

The spontaneity[15] of the act of discovery in the procedure of science has been long recognized. In fact such sudden appearances of novel ideas occur to all of us in much less pretentious situations. We say that the ideas "pop" into our heads, that we have "flashes of insight," often in the middle of the night, that we "suddenly recall" a name which had previously eluded us. Psychologists characterize it as an "Aha experience"; Darwin called it an "intellectual explosion"; Nicolle,[16] a "streak of lightning." Histories of science abound in illustrations (many of them doubtlessly spurious) of these sudden and unexpected discoveries. Archimedes discovered the principle of specific gravity while taking a bath; Watt discovered the principle of the steam engine while watching the tea kettle boiling on the stove; Poincaré discovered the fuchsian functions while boarding an omnibus; Gauss discovered the law of induction at seven o'clock one morning before arising. The case most frequently cited, of Newton discovering the principle of gravitation on seeing an apple fall from a tree, is probably spurious since Newton himself said that he discovered it by "thinking it over."[17]

One well-authenticated instance of this type of act may be cited for illustrative purposes, Sir William Hamilton's discovery of the quaternions (a mathematical concept concerned with vector quantities):

> They started into life, or light, full-grown on the 16th day of October, 1843, as I was walking with Lady Hamilton to Dublin, and came up to Brougham Bridge. That is to say, I then and there felt the galvanic circuit of thought *closed*, and the sparks which fell from it were the *fundamental equations between* I, J, K; *exactly such* as I have used them ever since. I pulled out, on the spot, a pocket-book, which still exists, and made an entry, on which, *at the very moment*, I felt that it might be worth my while to expend the labours of at least ten (or it might be fifteen) years to come. But then it is fair to say that this

---

[15] See references to footnote 11 (this chapter). All of these references abound in concrete illustrations drawn from science, both historical and current. Especially illuminating in view of the problem of how "scientific revolutions" occur is Thomas S. Kuhn, "The Structure of Scientific Revolutions," *International Encyclopedia of Unified Science*, II, No. 2, Chap. VI.

[16] Jacques Hadamard, *The Psychology of Invention in the Mathematical Field* (New York: Dover Publications, 1945), p. 19.

[17] Jevons, p. 581.

was because I felt a *problem* to have been at that moment *solved,* an intellectual *want relieved,* which had *haunted* me for at least *fifteen years before.*[18]

Helmholtz has given us an enlightening description of how scientific intuition occurs.

Often enough it steals quietly into one's thoughts and at first one does not appreciate its significance; it is only sometimes that another fortuitous circumstance helps one to recognize when, and under what conditions, it occurred to one; otherwise it is there, one knows not whence. In other cases it comes quite suddenly, without effort, like a flash of thought. So far as my experience goes it never comes to a wearied brain or at a writing-table. I must first have turned my problem over and over in all directions, till I can see its twists and windings in my mind's eye, and run through it freely, without writing it down; and it is never possible to get to this point without a long period of preliminary work. And then, when the consequent fatigue has been recovered from, there must be an hour of perfect bodily recuperation and peaceful comfort, before the kindly inspiration rewards one. Often it comes in the morning on waking up. . . . It came most readily, as I experienced at Heidelberg, when I went out to climb the wooded hills in sunny weather. The least trace of alcohol, however, sufficed to banish it. Such moments of fertile thought were truly gratifying.[19]

Some scientists argue that the spontaneous novelty of such acts can easily be over-emphasized. For example:

It is an erroneous impression, fostered by sensational popular biography, that scientific discovery is often made by inspiration—a sort of *coup de foudre*—from on high. This is rarely the case. Even Archimedes' sudden inspiration in the bathtub; Newton's experience in the apple orchard; Descartes' geometrical discoveries in his bed; Darwin's flash of lucidity on reading a passage in Malthus; and Einstein's brilliant solution of the Michelson puzzle in the patent office in Berne, were not messages out of the blue. They were the final coordinations, by minds of genius, of innumerable accumulated facts and impressions which lesser men could grasp only in their uncorrelated isolation, but which—by them—were seen in entirety and integrated into general principles. The scientist takes off from the manifold observations of predecessors, and shows his intelligence, if any, by his ability to discriminate between the important and the negligible, by selecting here and there the significant stepping-stones that will lead across the difficulties to new understanding. The one who places the last stone

[18] "Sir William Rowan Hamilton," in *North British Review,* XLV, 31.

[19] L. Konigsberger, *Hermann von Helmholtz* (Oxford: Clarendon Press, 1906), pp. 209–210. Reprinted by permission of the publisher.

and steps across to the terra firma of accomplished discovery gets all the credit. Only the initiated know and honor those who with patient integrity and devotion to exact observation have made the last step possible.[20]

Pierre Duhem argues in the same way.

History show us that no physical theory has ever been created out of whole cloth. The formation of any physical theory has always proceeded by a series of retouchings which from almost formless first sketches have gradually led the system to more finished states; and in each of these retouchings, the free initiative of the physicist has been counselled, maintained, guided, and sometimes absolutely dictated by the most diverse circumstances, by the opinions of men as well as by what the facts teach. A physical theory is not the sudden product of creation; it is the slow and progressive result of an evolution.[21]

Perhaps the contradiction between the opinions here expressed can be resolved by using the terminology of Wallas and arguing that the occurrence of "illumination" does not forbid the existence of a long, pre-inspirational period of "incubation," during which the scientist is "turning the problem over" in his mind. Certainly in all of the cases of discovery mentioned above, involving Archimedes, Watt, Poincaré, and the rest, there must have been much previous thought on the problem or the idea would never have arisen under any circumstances. The instances which are properly to be characterized as "serendipity" have the unique feature of being the occasions for the emergence of a new idea which had presumably never been investigated by the discoverer, and hence was completely unanticipated. Perhaps the most famous case of serendipity was Roentgen's discovery of X-rays.[22] Through mere chance some photographic plates, though protected from light, were found to have become fogged after being in the neighborhood of highly exhausted glass tubes or bulbs through which electric discharges had passed. Another example was the discovery of the voltaic cell. About the year 1786, Galvani noticed that the leg of a frog contracted under the influence of a discharge from an electric machine. Although Galvani attributed

[20] Hans Zinsser, quoted from *Scientific Monthly*, LXXX, No. 5 (May, 1955), 309. Reprinted by permission of the publisher.

[21] Pierre Duhem, *The Aim and Structure of Physical Theory*, trans. Philip P. Wiener (Princeton: Princeton University Press, 1954). Copyright 1954 by Princeton University Press. Reprinted by permission of the publisher. P. 221.

[22] W. C. D. Dampier-Whetham, *A History of Science* (Cambridge: Cambridge University Press, 1930), pp. 382–383. See also Kuhn, p. 57.

this to "animal electricity" it led to the invention of the voltaic pile.[23] Haüy discovered the structure of crystals when he accidentally dropped a piece of calc-spar upon a stone pavement. On examining the fragments he noted the regular geometrical faces, which did not correspond with the external facets of the crystals.[24] The commonly cited example of Simpson's discovery of the anaesthetic power of chloroform when a saucer containing some of the liquid had a strange effect on a dog who happened to smell it, has been shown to be spurious,[25] since Simpson already knew about chloroform and had actually been experimenting with it previous to the supposed incident.[26]

6. Aids to Insight

Realizing, then, that there is probably no sense in which one can say strictly that the scientist can prepare for the act of illumination (other than by familiarizing himself as completely as possible with the relevant data), nevertheless there are certain methods of operating upon the data which do seem to foster the emergence of novel ideas. One of these, which we have already examined in detail, is measurement. We have seen that this is a process by which, using established rules, we attach numerals to objects and events. The result is a system of symbols, but the symbols are numerals, not words, and the system of such symbols is a mathematical system and not a group of sentences and paragraphs. The importance of mathematics in fostering scientific discovery lies in the fact that, among all the sciences, it has achieved the highest degree of logical organization. If the reader can recall his high school geometry he will remember that he had to begin by learning certain definitions, axioms, and postulates. These told him what points, lines, and surfaces were, and informed him that lines could be extended indefinitely, that figures could be moved about without distorting them, that quantities could be added and subtracted according to certain specified rules, and so on. Then he learned certain theorems, all of which could be logically proved by deriving them from the definitions, axioms, and postulates, and

[23] Dampier-Whetham, p. 232.
[24] Jevons, p. 529.
[25] *Ibid.*, p. 532.
[26] For further illustrations see Beveridge, Chap. III and Appendix.

he was told that he would "know" geometry if he could carry out this proof in the case of each theorem. The neatness of the geometric scheme was indicated by the fact that there were no "dangling" elements—no notions which did not fit into the scheme, and no theorems which were incapable of such proof. Arithmetic, algebra, and other fields of mathematics have achieved this same type of logical order.

What happens when we use measurement techniques in descriptive science? As we have seen, we set up a method by which we can attach to each object a number or a group of numbers corresponding to its quality or qualities. What is involved in all measurement is the employment of a system of numbers to represent a system of qualities. Qualities which have been measured become quantities; for each number there is one quantity and for each quantity there is one number. Systems of this kind are said to have a one-to-one relation to one another. By this is meant that the elements of the one system correspond to the elements of the other, and the relations of the one system correspond to the relations of the other. Now the important point to be noted here is that the aggregate of numbers is itself a deductive scheme. Consequently, when we apply this scheme to our qualities these also take on the deductive order. Any relation which is then known to hold among the quantities may be assumed, with certain restrictions which need not be mentioned here, to hold among the qualities. Thus we have an important technique for discovering further relations among our facts. A few examples will suffice to make our point clear. Any two numbers may be multiplied together and a new number produced. Similarly two lengths may be multiplied together and a new quantity, an "area," is produced; mass and acceleration may be multiplied to produce a force, and volume and density may be multiplied to produce mass. One number may be divided by another. Similarly a change in space may be divided by a lapse of time and the result is a velocity. Mathematical quantities are related in a manner called "functional" when for each value of the one there is at least one value of the other. In the same way qualities which are measured by numbers become related functionally. These are all cases in which the qualitative aspects of our experience become integrated by the application of the mathematical scheme to them.

Unquestionably the most important device for systematizing

the facts of a descriptive science is *idealization,* or "passing to limits." Eddington calls it the method of "just like this only more so." It is an elaborative and refining process by which we try to get away from the complexities and crudities of observed objects in order to replace them by simpler and more precise counterparts. In order to realize our ideal we must idealize our real. One of the best examples is the notion of a perfect lever. The simplest lever is an iron bar allowed to move freely about a pivot. The bar can be balanced at approximately its mid-point and is then said to be in equilibrium. If equal weights are attached to the ends, the bar will remain in equilibrium; but if a heavier weight is added to one end than to the other, the pivot must be shifted so as to shorten the arm to which the heavier weight is attached. Proceeding in this way and introducing measured values, we can determine a simple relationship between the weight and the length of the arm for conditions of equilibrium. This is the law of the lever. It states that the product of the weight by the length of the arm is a constant quantity. Now in a crude experimental situation, such as we have just considered, the results would not be in exact accord with this law; they would vary, much as the actual boiling point of water fluctuates around 212°. The lengths cannot be precisely measured; the bar does not move with perfect freedom on the pivot; when the weights are very great the bar may bend considerably; the bar is probably not itself exactly homogeneous so that one end may weigh more than the other; and so on. When the scientist is confronted with an occasion of this kind he invents the notion of an "ideal situation" in which there would be no disturbing factors; then he asserts that in such a perfect case the law holds exactly.

The frequency with which such notions occur in science is not generally recognized. They are found not only in the physical sciences, but in the biological and behavioral sciences as well. They are very common in mathematics itself, particularly in geometry, where lines have no breadth and thickness, circles are perfect, and the sum of the angles of the ideal Euclidean triangle always equals exactly 180°. Substances which are chemically absolutely pure, and gases which exactly obey the gas law are common conceptions in chemistry. In the social sciences frequent use is made of such fictions as the completely isolated individual, the ideal State, and the purely economic man. *Averages,* while they are not really idealizations, do frequently correspond to nothing in experience. We may know that the average size family in a certain community is 3.2 persons, though we should have difficulty finding *this* family; and we may know that since some human beings have only leg, while few if any have more than two, the "average" man must have 1.99999 . . . legs—but we do not search for *this* man.

Terminology which is suitable to describe these symbols is hard to find.[27] They have been variously called "constructs;" "fictions," "idealizations," "abstractions," "archetypes," "as-if notions," "counterfeit ideas," "limiting concepts," "utopian ideas," "artifices," and "inventions." All of these expressions endeavor to convey the double role which these "entities" play—their great utility in science, yet their non-representative (in the ordinary sense of the word) relation to the objects of everyday experience. Einstein felt that this separation of symbols from the world was important for science. He said, "I consider it incorrect to veil the logical independence of (scientific) concepts from sense experience. The relation is not analogous to that of soup to beef, but rather that of the check-number to the overcoat . . . deposited in the checkroom."[28]

The motive which is responsible for the idealizing techniques is the desire to explain. Perfect levers are to be preferred to actual levers because they are simpler, their laws can be stated exactly, and they can be integrated and systematized with other scientific notions more effectively. Implicative relations do not hold among fuzzy symbols. Similarly in an ideal State we do not need to bother our heads about graft, power politics, and pork barrels, since everyone will be working for the good of the group. Hence we can neglect these disturbing factors and construct the notion of a State in which the theoretical ideas of political organization are perfectly exemplified. Logical neatness in our symbolic scheme is thus achieved and its subdivision into the derived and the underived symbols becomes considerably simplified. As a conse-

[27] The most comprehensive development of the notion of *fictions* is to be found in Hans Vaihinger, *The Philosophy of 'As If'* (New York: Harcourt, Brace & Co., 1925). See also C. K. Ogden, *Bentham's Theory of Fictions* (Paterson, N.J.: Littlefield, Adams & Co., 1959); Morris Cohen and Ernest Nagel, *An Introduction to Logic and Scientific Method* (New York: Harcourt, Brace & Co., 1934), pp. 367–375; and Israel Scheffler, *The Anatomy of Inquiry* (New York: Alfred A. Knopf, Inc., 1963), pp. 185–222.

[28] A. Einstein, "Physik und Realität," *Zeitschrift für freie deutsche Forschung* (Paris, 1938), p. 9.

quence the way is prepared for the discovery of illuminating and unifying hypotheses.

With the achievement of this result there is, however, an unfortunate consequence. Descriptive accuracy is to a certain extent lost. The world does not contain perfect levers, ideal gases, purely economic men, and utopias; when we build science on such notions, therefore, we are no longer talking about the world in which we live. Our science ceases to be applicable, except in an approximate sense. Many who praise the exactness and precision of science fail to realize that only through the use of such fictions and idealizations does science have its mathematical rigor. Applied science fits the world better, but it lacks the neatness and organization of idealized science; idealized science has the required coherence, but it loses to a great extent its descriptive value. In the process of idealization we can see clearly how the desire to explain so modifies the descriptive science that it begins to lose its contact with fact. The desire for neatness and logical system prevails over the desire for faithfulness to the facts.

Thus far we have discussed two of the techniques by which we prepare a descriptive science for that inductive leap which will transform it into an explanatory science. There are other procedures which may be mentioned in passing. Analogy, for example, is a convenient device for getting explanatory conceptions. This is a technique by which we compare two objects, one of which is better known than the other, in order to learn more about the less known object. To do this we examine the two objects, point by point, noting all the qualities which they possess in common; then we argue that since the objects resemble one another in so many respects they probably are alike in further qualities. We then feel we are justified in supposing that an additional property which is possessed by the object which we know, will also be possessed by the one which we do not know. Suppose, for example, knowing that sound is transmitted by waves in the air, we ask how light is transmitted. We note that the intensity of light varies with the distance from the source, as it does in the case of sound; that the color of light changes if the source is moving rapidly toward or away from the receiving point as does the pitch of sound; and that the character of a light stimulus is changed if it is made to pass through water rather than air, as is also true of sound. We then conclude by analogy that light travels in waves as does sound.

This is not, of course, proof; analogy never proves anything. But explanatory hypotheses which have been suggested by analogy can often be confirmed by the deductive techniques which we shall examine later. There are principles which guide analogy. For example, in general the more qualities two objects have in common the more likely it is that they will possess a further property in common. But this and other principles do not tell us how to pick out significant analogies. Hence, while analogy is an aid to insight it does not really *produce* this important quality of mind.[29]

Another aid to insight is generalization. Suppose in a given descriptive science there is a series of statements about conductors of electricity; we have examined many substances and have found that iron, gold, silver, copper, and tin, all of which are metals, conduct electricity. If we are not positivists we shall ask ourselves why this is the case. The asking of this question is equivalent to searching for an hypothesis from which it will be possible to deduce the statements as logical consequences. A little ingenuity, aided by the rules of logic, tells us that if there were a law stating that *all* metals conduct electricity we could derive not only each of these statements but some additional ones applying to metals which we have not yet examined. We could not only explain why these substances conduct electricity but we could provide ourselves with advance knowledge of what we should discover if we were to look further into nature. But can we guarantee such a law on the basis of the cases actually examined? Evidently not, for, as we have seen, our "movement of thought" is from "implied" to "implies," not from "implies" to "implied"; the best we can expect, therefore, is a certain degree of probability. If we want a higher degree of confirmation we shall be obliged, as in the case of the explanation of light, to resort to deductive techniques. To be sure, there are some aids to this process of generalization: we should have a large rather than a small number of cases to start with, and the instances should be "representative" rather than unusual or abnormal. But this cannot be the whole story; the essence of the inductive leap is again to be found in insight. Why, for example, can we sometimes generalize with a high degree of probability on the basis of only a few instances, and on

[29] For a number of scientific examples of the use of analogy see Jevons, Chap. XXVIII.

other occasions only with a great risk of error even though we have many cases?[30] We can justifiably pass almost immediately from "one book plus one book equals two books" and "one stone plus one stone equals two stones" to "$1 + 1 = 2$"—as children frequently do—and be quite certain of the result. But we should not forget the rooster who concluded that since he had been fed at sunrise every day of his life this would always be the case, only to find that on a certain morning his neck was wrung instead.

Beyond these few techniques there is little that can be suggested as a guide to induction. In its essential nature the movement remains a mystery. That it has close affiliations with imagination we shall see when we return to the subject in Chapter 8. Kekulé, a nineteenth-century chemist, said, "Let us learn to dream, gentlemen, then perhaps we shall find the truth."[31] And Whitehead asserted that under the impetus of the scientific imagination, "a fact is no longer a bare fact: it is invested with all its logical possibilities. It is no longer a burden on the memory: it is energizing as the poet of our dreams, and as the architect of our purpose."[32]

We have noted certain conditions under which the flash of insight frequently occurs—Archimedes while taking a bath, Watt while watching a kettle boil, and Gauss while lying in bed at seven o'clock in the morning. But it would be absurd to formulate certain rules stating that the way to make discoveries is to take more baths, watch more teakettles, and awake every morning at seven. We learn when we read the reports of men who have had such flashes of insight that these seldom occur when the mind is tired, or even when it is hard at work, but rather in periods of relaxation; this urges upon us the importance of occasional rests, but does not guarantee that in idle moments we shall always have flashes of insight. Scientific biographies inform us that geniuses are always driven by powerful urges; but the strength of the desire for an insight is never a sufficient condition for its occurrence. All that we can do, therefore, concerning this important aspect

[30] Mill states that "Whoever can answer this question knows more of the philosophy of logic than the wisest of the ancients, and has solved the problem of induction." J. S. Mill, *System of Logic* (8th ed.; New York: Longmans, Green & Company, 1911), Book III, Chap. III, Sec. 3, p. 206.

[31] Quoted from Bernard Barber, *Science and the Social Order* (Chicago: The Free Press of Glencoe, 1952), p. 203.

[32] A. N. Whitehead, "Universities and their Function," *Atlantic Monthly*, Vol. 141, May 1928, 639.

of science is to list the conditions in the absence of which it seldom occurs; we can state its necessary but not its sufficient conditions. We can say, simply, that unless the scientist has a firm foundation of extensive factual knowledge, unless he has an overpowering drive, unless he indulges in occasional relaxation, unless he understands something of the logic of explanation—he cannot hope for results. But if he satisfies all these conditions he may still not succeed if he lacks the mysterious spark. This is the bitter conclusion to which we are driven.

FIGURE 8

DESCRIPTIVE SCIENCE AUGMENTED BY INSIGHT AND INDUCTION

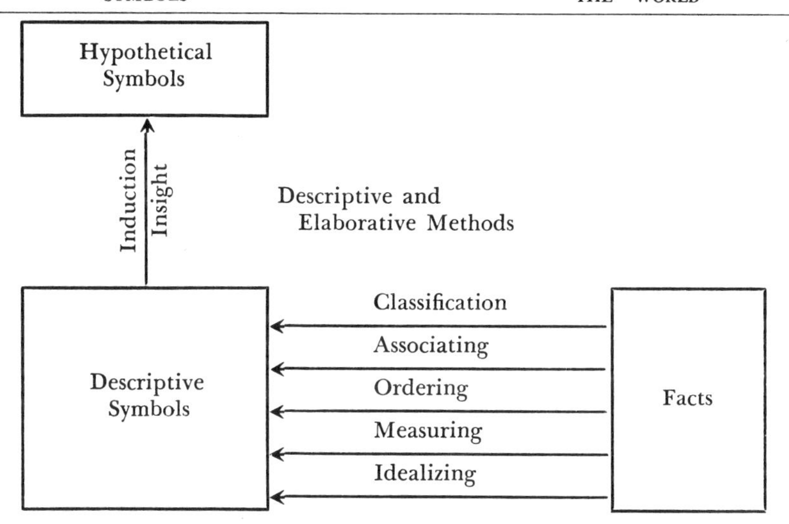

We are now ready to summarize the results of this chapter and to indicate by means of a new diagram how a descriptive science is extended through the act of discovery. We can readily see that our science is now more inclusive than it was, since we have introduced, through insight and induction, various types of objects which have never been disclosed through observation, and which, so far as we know at present, may not exist at all. We cannot, therefore, put such theoretical objects in the realm of fact. Only by means of the confirmatory processes, which we shall examine in the next chapter, can we determine whether these entities really exist or are mere figments of the imagination. Furthermore, while this more expanded type of science still retains its descriptive features, it has yielded to the demands of explanation and has introduced idealized and perfect entities of various kinds. It has, thus, to a certain extent lost the directness of its descriptive representation. In this respect it has not broken completely with purely descriptive science, since, as we saw at the end of the last chapter, even the simple descriptive processes of observing, manipulating, interpreting reports, and describing involve some loss in the directness of the representation.

The diagram in Figure 8 indicates the kind of science we have when description has been augmented by tentative explanation, but when the testing and verifying techniques have not yet been employed.

How Theories Explain

We have seen that when descriptive techniques have been supplemented by insight the result is a very unstable science, one which has ceased to be descriptive but is not yet truly explanatory. We believe that if the facts which we have collected and described are to be explained, this explanation will probably come through some idea that has been given to us by insight. We have a hunch that the world is more than we have discovered through observation, but we have no guaranty that this is the case. We know that insight is unreliable. Hence we are in the unfortunate position of having made what seems like a good guess without being able to assure ourselves that it is anything more than this.

A noteworthy characteristic of flashes of insight, if we are to accept the reports of scientists who have had them, is that they almost always produce in the minds of their originators an overwhelming conviction of truth. Not only does the idea itself appear suddenly, but the whole set of the mind is instantly changed from one of puzzlement, perplexity, and tension, to one characterized by that peace of mind which is found only when problems have been solved and difficulties overcome. In this respect insights are very much like the hunches which occur to all of us in less pretentious situations. Frequently we experience a sudden dislike for a person whom we have just met—a feeling which is often borne out by later acquaintance with the individual; sometimes we have premonitions that something unfortunate for ourselves or friends is going to happen; sometimes, after thinking a long time about a business deal or a complicated moral matter which we have been unable to resolve because of a conflict of principles, we have a sudden illumination, perhaps in the middle of the night, and the problem is solved in a flash. Women are supposed to be particularly blessed with this power of intuition. In all such cases there seems to be associated with the insight a strong feeling that we cannot possibly be mistaken; we seem to have had a direct vision of truth, and we immediately discount any evidence to the contrary since it seems pale in comparison with the overwhelming conviction which is attached to the insight itself. We proceed at once to act on the basis of our hunches, and feel no misgivings that we might be making a mistake. When we have intuitions of the character of another person this often involves modifying our behavior toward that individual; when we have premonitions we frequently change our plans, in order, as we say, not to tempt fate; when we have problems which are suddenly solved we promptly act on the basis of our intuitions and drop the problems from our minds because we have experienced that relief from tension which is always found when a decision has been reached.

The tragedy is that in so many cases both hunches and scientific insights are wrong. Firmness of conviction is no guaranty of truth.

> The first condition to be fulfilled by men of science, applying themselves to the investigation of natural phenomena, is to maintain absolute freedom of mind, based on philosophic doubt. Yet we must not be in the least sceptical; we must believe in science . . . we must believe in a complete and necessary relation between things . . . but at the same time we must be thoroughly convinced that we know this relation only in a more or less approximate way, and that the theories we hold are far from embodying changeless truths.[1]

Most people who have had sudden insights find it difficult to doubt them, since they remember the cases in which their hunches

---

[1] Claude Bernard, *Introduction to the Study of Experimental Medicine* (New York: Dover Publications, 1957), p. 35.

turned out to be correct and forget those in which they fell wide of the mark. There may be people who are more often right than wrong on such matters, but they are certainly few in number, and even such individuals have no infallible guaranty that their hunches are correct. The batting average of scientists on the matter of insights cannot be computed, for investigators do not as a rule record for posterity the cases in which they made bad guesses. When we read scientific biographies and are confronted with report after report of sudden insights which were later corroborated by the usual verificatory processes we are tempted to conclude that the scientific mind is particularly skillful at this sort of thing, and that in such cases the mere occurrence of an illumination is a certificate as to its truth. But if we could have a history of scientific failures as well as scientific successes we might draw a different conclusion. Darwin said, "I cannot think of a single first formed hypothesis which had not after a time to be given up or greatly modified." [2]

Because hunches are not self-certifiable we must resort to some other technique for establishing their validity. We shall examine first a method which is applicable in certain restricted situations.

1. Direct Testing

Let us suppose that on a certain cold morning your car will not start; perhaps the battery will turn the motor over but it will not "take." This gives rise not only to a simple descriptive science but to a descriptive science which cries out for extension into the realm of explanation. The factual proposition is that all of the usual conditions for starting the motor have been met, yet it will not start. The invitation to go on into the realm of explanation is indicated by the persistence of the question, *Why*? To be sure, the urgency of the problem may be primarily one of practical behavior since you presumably have an appointment which you will be unable to keep if the car will not start. But regardless of why you want to answer the question, there is the question itself which demands an answer. If you know anything about cars you will

[2] Quoted by W. I. B. Beveridge, *The Art of Scientific Investigation* (New York: Random House, Inc., 1957), p. 79. Beveridge adds, "W. H. George points out even with men of genius with whom the birthrate of hypotheses is very high, it only just manages to exceed the death rate." P. 80.

immediately find yourself making guesses as to what is wrong. You may be out of gas, the battery may be too weak to produce a good spark, or the distributor points may be corroded. Each of these hypotheses which occurs to you is an instance of scientific illumination functioning in a commonplace situation, and you are proceeding exactly as a scientist would in such a case. Let us suppose you have a hunch that the gas tank is empty, and you favor this hypothesis above the other two. Since you know that such intuitions are unreliable you resort to the simple act of looking at the gauge. If when the ignition is on you find that the gauge registers "empty" you have verified your hypothesis by the most direct route possible. (Many unexpressed assumptions are, of course, here involved.) It is no longer a *guess* but a *fact* that you are out of gas. By the simple process of looking for the fact which is asserted in your hypothesis you establish its truth. Indeed, you use exactly the same method, except for the direction of your activity, which you employed when you learned that the motor would not start. You had the idea of the empty gas tank *before* you discovered that fact, but you had the idea of the failure of the motor to start only *after* you observed it. What you have really done, therefore, is to anticipate certain facts on the basis of what you knew about other facts.

Most situations of popular reflection involve direct testing of this kind. We may refer again to the procedure of the detective in solving a mystery. Confronted with the corpse, or the fact of missing jewels, or some other evidence of crime, he proceeds to induce a number of theories as to who the culprit is. Usually the disproof of all but one of these theories is a complicated and involved process based upon the use of deductive techniques which we shall examine in a few minutes. Frequently, however, the criminal is compelled at the end of the story, by the force of accumulated evidence, to make a confession. Such corroboration is of the same direct kind as that involved in determining whether the gas tank is empty. Most situations in which applied science is used for the solutions of problems proceed along the same line, whether it is a matter of replacing a broken wire when your doorbell will not work, curing a case of appendicitis by removing the appendix, or relieving a youth of his criminal tendencies by changing the character of his home life. In all such situations the facts lead us to formulate theories as to causes, which we then proceed to test by finding (or not finding) these causes. That we wish to remove the causes in order to remove the effects is purely incidental so far as the problem of understanding is concerned. These illustrations show that very often the process of testing consists in a simple comparison of the hypothetical idea with a directly observed, or immediately reported, fact. When such occurs the theory ceases to be conjectural and becomes descriptive of fact, and we then have an expanded science in which there are not only statements descriptive of effects but also statements descriptive of causes.

In many practical situations, however, and in the great majority of significant scientific situations direct corroboration of hypotheses and theories is not possible. This is the case whenever the fact which is asserted in our theory is such as not to be observable. There are many reasons why this may be the case. The fact, for example, may have occurred in the past and may no longer exist; since we cannot turn time backward and relive the past we have to admit that it is gone forever. Historical sciences, therefore, can never use the method of direct corroboration, for even in the case of the report of a witness the existence of doubt as to the authenticity of the report and the accuracy of the witness proves that the verification is not really of a direct sort at all. The indirect evidence may be overwhelmingly in favor of a certain interpretation, but since the historian cannot himself experience the event his belief in its occurrence can be only a matter of probability. Other facts are unobservable for different reasons. Because the moon always turns only one of its faces toward us, we cannot observe its other side; consequently, any hypothesis as to its nature must receive indirect rather than direct corroboration. The Russian scientists who were successful, if they *were* successful, in sending a rocket around the moon produced direct, or at least photographic, evidence as to its nature. Any hypothesis concerning the nature of the center of the earth must today receive only indirect corroboration. Many things, such as atoms and electrons, are too small to be seen even with the aid of instruments. Other things are too remote to be seen. Still others, such as the earth, are too large, though here again we have photographic evidence and there may soon come a day when at least one hemisphere of the earth will be seen by man. The same restrictions in observation occur because of the limitations in the other senses. Again, any knowledge I may have concerning the nature of your mind must remain, for me, always indirect. True, I may talk with you and ask you to tell me something about "what is going on in your head"; but this, again, is an indirect report and because of the difficulties in communicating it may be very misleading. Hence my knowledge of your mind may be wholly mistaken and I can never check it in the same way that I can verify knowledge of my own mind. Some have proposed that there may come a day when the brain of one individual may be grafted by means of nerve fibres to the brain of another; if this ever occurs direct awareness of the consciousness of another individual may become an actuality.

The conclusion to which we are driven, therefore, in connection with all such unobservables is that if they exist we can determine this not by direct inspection of the facts themselves, but only by discovering other facts to which they are intimately related. Strictly speaking, the only thing which exists for an individual is something which he has himself observed. But he soon learns to accept the reports of others concerning what they have observed, and to include these objects within the field of what he considers to be real. In the same way, with the refinement of observational techniques and the use of instrumental aids, he finds himself forced to admit that there are many objects in the world which he had not previously known to be there. This leads to the very important generalization that the world probably contains more than is at any stage in the development of knowledge actually observed. An individual might construct a list of the objects of the world, arranged in the order of the certainty with which their existence can be affirmed. Such a list could be something like the following:

A. Objects which the individual is now observing.
B. Objects whose existence is reported by instruments.
C. Objects whose existence is reported by the words of other individuals.
D. Objects which he has observed in the past.
E. Objects which he will observe in the future.
F. Objects whose existence he can never know through direct observation or through reports, but whose existence is demanded for the purpose of explaining other objects which he does know to exist.

For the average scientist facts of the kinds A, B, C, and D are included in descriptive science. Although there is, among these four items in the series, an order of decreasing probability concerning the reliability of our information, nevertheless we usually consider all such objects to be real, and we endeavor to build our science on them. Objects of the kinds E and F, however, have only a conjectural status; they exist as hypothetical objects whose reality can be determined only by indirect methods. In the diagram given on page 116 all objects of the kinds A-D would be located not only in the realm of facts but within the special area containing those facts which have been collected by the methods of observation and report. Objects of the kind E and the kind F cannot, however, be put into the realm of fact at all; they exist as theories, not as facts. We must put them into the area under "symbols," since they are theoretical notions—notions which are still under debate and which cannot be placed in the realm of fact until their existence has been verified through deductive techniques.

2. Deductive Verification

We are now ready to see what is involved in the method of deductive confirmation.[3] Let us begin with a good working definition of a hypothesis. (We shall continue to use the words "hypothesis" and "theory" interchangeably. Often, as was suggested earlier,[4] a distinction is made in terms of the degree of confirmation al-

[3] The material to be discussed in the next two sections is commonly treated, with varying degrees of success, in the inductive portions of the general textbooks in logic used in colleges and universities today. My main criticism of these books is that while they have modernized their sections on deductive logic because of the rapid developments in symbolic logic in recent years, they have not taken cognizance of the equally important improvements in our understanding of the scientific method. Two books which suffer least from this inadequacy are Morris Cohen and Ernest Nagel, *An Introduction to Logic and Scientific Method* (New York: Harcourt, Brace & Co., 1934), and H. A. Larrabec, *Reliable Knowledge* (Boston: Houghton Mifflin Company, 1945), already referred to. Other more comprehensive books are the following: Ernest Nagel, *The Structure of Science* (New York: Harcourt, Brace & World, Inc., 1961); Arthur Pap, *An Introduction to the Philosophy of Science* (Chicago: The Free Press of Glencoe, 1962); Israel Scheffler, *The Anatomy of Inquiry* (New York: Alfred A. Knopf, Inc., 1963); Norwood R. Hanson, *Patterns of Discovery* (Cambridge: Cambridge University Press, 1958); and certain collections of readings, such as *Philosophy of Science,* ed. Arthur Danto and Sidney Morgenbesser (New York: Meridian Books, Inc., 1960), and *Readings in the Philosophy of Science,* ed. H. Feigl and M. Brodbeck (New York: Appleton-Century-Crofts, Inc., 1953).

[4] See above, p. 26.

ready achieved, a hypothesis being a mere guess, while a theory has a certain established status. Actual usage in science provides no justification for this relative distinction, and we shall use the words as synonyms.) Bavink defines a hypothesis as "the presumption of the existence of a general state of affairs, lying at the back of certain phenomena which are matters of experience, and allowing the phenomena in the field of facts in question to be deduced qualitatively and quantitatively (mathematically) from the presence of said state of affairs and its assumed laws."[5] The method of deductive confirmation is that by which the existence of the theoretical entities or states of affairs is established. It always takes the following form: If the statements about the theoretical objects are true there should be true statements about certain facts still to be observed. If, for example, there are molecules, then when certain solids are put into certain liquids and the mixture heated or stirred a homogeneous solution must result. We perform the simple experiment with salt and water, and note that the expected consequences occur. We then conclude (with a certain probability) that molecules exist. The steps in this process are two in number. First, there is the deductive movement, the movement of prediction. Second, there is the testing movement—the movement of confirmation or disconfirmation, the movement in which the predicted consequences are compared with the actual situation in the world. Let us examine briefly these two important movements.

The deductive movement always takes the "if—then—" form, where what immediately follows the "if" is the theoretical notion whose existence we are trying to establish. This is the hypothesis which was suggested through insight or inspiration and whose truth seems likely simply because of the conviction which is ordinarily associated with such flashes of illumination. But since we have learned that these hunches are not to be relied upon we undertake to establish their probability by some more objective means. We should like, if it were at all possible, to observe directly the facts which are asserted in our hypothesis. But we can do this only very rarely in science. Usually if they exist at all they occur beyond the range of observation and their existence must be determined indirectly. The movement of prediction thus in-

[5] B. Bavink, *The Natural Sciences* (New York: The Century Co., 1932), p. 42. Reprinted by permission.

volves the analysis of the hypothetical notion to see what would be the consequence of supposing that its object existed. We simply say to ourselves: Let us imagine that there were such a thing. How would it manifest itself in terms of concrete differences in observable phenomena? What sort of world would this be, for example, if there were molecules? By analysis of the notion of molecules and molecular motion we can anticipate how such entities "ought" to make their presence known. If the world proves to be this kind of place rather than some other, we may conclude on rational grounds that molecules probably do exist. We say that the existence of molecules implies the existence of solutions, e.g., salt dissolved in water, or that the existence of solutions can be logically deduced from the existence of molecules.

The exact nature of this movement of thought can be readily seen by comparison with the inductive movement, or the movement of discovery. As we saw in Chapter 5, one is exactly the reverse of the other. In the inductive movement we go from fact to theory; in the deductive movement we go from theory to fact, or at least to supposed fact. This formulation is, of course, not strictly correct, as will be seen later from the diagram on page 133. In induction we go not from fact to theory but from a statement which is a true description of a fact to a statement expressing a theory; similarly in the deductive movement, as thus far analyzed, we go from a statement of a theory not to a fact but to a statement which (we hope) will prove to be a true description of fact. (This is simply to say that we cannot define deduction and induction as relations between facts, because what we normally have when we perform these logical operations are not the facts themselves but only statements which actually may or may not be factual.) As we have already noted the theory is not necessarily contained in the description, since other theories might do the explaining job equally well. But the presumed description of the still-to-be-discovered data is necessarily contained in the theory.

The other step in the complete movement of testing involves nothing new. Indeed, it is simply the reverse of the movement by which truths are established in a descriptive science. Thus the techniques of observation, accepting instrumental and verbal reports, experimentation, classification, association, ordering, measurement, and idealization may all be involved. But now, instead of accepting the facts as given and then proceeding to their accurate description or idealization, we start with the description or idealization of certain supposed facts and proceed to determine whether the predicted facts really exist. Instead of going from "facts" to "ideas" we go from "ideas" to "facts." In the one we discover facts and then classify, associate, order, measure, and idealize them; in the other we start with the classes, associations, orders, measured values, and idealizations, and endeavor to find facts which will be members of the classes, cases of the associations, instances of the orders, examples of the qualities to which the measured values can be attached, and approximations to the idealizations which have been created.

Because confirmation is essentially the reverse of description, much that was said in Chapter 2 and 3 applies equally well in this connection. But there are some noteworthy differences. Attention, for example, is much more acute in the testing stage than in the stages of mere collection and description. The observer is now *looking for* something specific; he *knows* what he is looking for. In the descriptive stage he is more or less passively waiting for something which appears to be sufficiently interesting to arouse his curiosity; he is, as we have already seen,[6] merely *listening* to nature. But in the testing stage he has a definite idea in mind; he is expecting one sort of occurrence rather than another and he is thus in a much better position to notice this event if it is really there. He is now *asking nature questions.* Merely to record what takes place without any advance suggestion as to what to look for, or what to expect, is a very inefficient process. Many an object or event escapes our notice completely, either because we have not yet had a sufficiently broad background to interpret what is happening and thus make it part of our experience, or simply because we are not attending to it. There are marked differences between what the botanist, the artist, and the common man see in a flower; each selects attributes according to his own interests. In much observation one sees only what he looks for. And since what he looks for is dependent on individual interests and past experiences, what is actually seen varies from person to person. In spite of these changing factors, however, the trained observer—the one who knows what to look for—can always see more than the untrained observer.

[6] See above, p. 34.

But, on the other hand, observation which is carried on under the guidance of a definite purpose is subject to corresponding dangers. If we are looking for something specific we are not only much more likely to overlook other facts but — what is more serious — we are almost certain to misinterpret what we do observe so as to make it agree with our theory. If we are strongly hopeful that a certain event will occur, we may believe that it *does* even though this is not really the case at all.[7] An example from my own experience may indicate what is involved here. A number of years ago — such a large number that I hesitate to specify it — I was driving an automobile whose tires, unfortunately, were in very bad condition. On the particular day when the event occurred I had already had two punctures, and I knew that the other tires were badly worn. As I was driving along, with the set of my mind obviously determined by what had already happened, I heard an explosion. Whether it was a stone hitting the fender or a backfire from the engine I shall never know. But at the time it meant only one thing for me — blowout. Suspecting the left rear tire, I thrust my head out of the window as I was slowing down, and noted distinctly the bulge of the tire on the pavement. I also felt the "thump-thump" of the valve-stem. I was using at the time a so-called puncture-proof liquid which was supposed to flow into any hole in the inner tube and prevent the tire, at least temporarily, from deflating (modern tubeless tires take care of this more efficiently today). This liquid had a peculiar smell, and I definitely noted the odor of this liquid as it escaped through the hole in the tube. After stopping I proceeded immediately, before examining the tire, to get out my jack and repair kit, which, in those days, were carried under the front seat. I was astonished to find that the tire and, in fact, each tire on the car was properly inflated. I had been deceived by three of my senses — I had felt the vibration of the valve stem, I had seen the bulge of the tire on the pavement, and I had smelled the liquid — when none of these "facts" existed. I shall always believe that there was an explosion of some sort, but whatever it was it was sufficient to cloud my observational powers to such an extent that I was completely unable to get a true picture of the situation. This is, of course, an extreme case. But milder instances of the same kind

[7] Mill speaks of "The deep slumber of a decided opinion." J. S. Mill, *Utilitarianism, Liberty and Representative Government* (London: J. M. Dent & Sons, 1910), p. 103.

will occur to almost anyone who has been at all critical in his observations. In science the danger lies particularly in connection with theories which have already received a certain degree of confirmation and to which the scientist has thus developed a certain emotional attachment. He has great difficulty in convincing himself that he is observing a fact which refutes the theory, and he therefore misinterprets the fact so as to bring it into agreement with his theory.

There is another important difference between the original descriptive movement of science and the testing process which we are now examining. This concerns the manipulatory acts which we introduce in our desire to prod nature into speaking. Whereas in the descriptive stage we have no advance knowledge of what nature is going to tell us, in the testing stage we have such information—information which has been derived from the theory whose truth we are trying to ascertain. These controlled or guided experiments are commonly called, as we have seen,[8] "experiments for confirmation," in contrast to the uncontrolled and unguided experiments of the earlier stage which are called "experiments for discovery." The latter type never produce disappointments; for since there is no advance knowledge of what is going to occur there can never be surprises of any kind. The testing experiments, on the other hand, often fail to occur as anticipated and we are then compelled to revise our theory.

3. Confirmation and Disconfirmation

Let us now examine what steps we are entitled to take in the two possible cases of testing—when the observational or experimental results *correspond* to the predictions of our theory, and when they *fail to correspond*. In the former case we say that the theory has been confirmed. But we should note that the testing of a single consequence of the theory constitutes only a part—indeed, a very small part—of the desired confirmation. We must accumulate evidence over a wide area before we can assert that the theory has a very high degree of probability. In general the probability of a theory depends on the *number* of confirmed consequences, their *variety and apparent unrelatedness*, and their *unexpectedness*

[8] See above, p. 34.

without the supposition of the theory they are designed to test. The explaining power of the theory lies precisely in this capacity to unify a wide range of otherwise unrelated facts. The power of the electromagnetic theory lies in its capacity to bring light and electricity together into a single unifying conception; the theories of gravitation, of evolution, of heredity, the heliocentric theory of the heavens—all these gain their value as explanatory notions simply because of the wideness of their sweep, and because of the number and variety of predictions from them which have been and are still being confirmed by observation and experimentation. To refer again to detective stories, the task of the detective-hero, and therefore also of the reader, is to accumulate more and more evidence in support of the guilt of a certain suspected individual, until it becomes so overwhelming that a jury could pass a verdict upon him "beyond any reasonable doubt," as the phrase has it. This involves showing that he had the motive and the opportunity, that the evidence at the scene of the crime is such as we should expect if he had committed the act, that his behavior before and after the crime were consistent with his having done so, and so on. One swallow does not make a summer, nor does one verified consequence completely confirm a theory.

The procedure in the case of the disproof or disconfirmation of a theory would seem to be much simpler. For whereas the confirmation of a hundred predictions from a hypothesis would do no more than render it highly probable, a single disconfirmation would disprove it completely. The character of the implicative relation between P and any of its deduced consequences is such that if P implies Q,R,S, and T, all of which are tested affirmatively, we can attribute to P only a certain degree of probability; but if *any* of the predictions, Q,R,S, or T proves not to occur as anticipated, then P must be rejected. For if P implies Q, the falsity of Q implies the falsity of P. Thus it is commonly said that the disconfirmation of a hypothesis is much simpler than its confirmation. Perhaps this is what led Hermann Weyl to state, "Once and for all I wish to record my unbounded admiration for the work of the experimenter in his struggle to wrest *interpretable facts* from an unyielding nature who knows so well how to meet our theories with a decisive *No*—or with an inaudible *Yes.*"[9]

[9] Hermann Weyl, *Theory of Groups and Quantum Mechanics* (London: Methuen & Co., Ltd., 1931), p. xx.

This character of the scientific method does, however, suggest a manner of proceeding when a number of alternative hypotheses is found to explain a given object or event. If it should be the case that M,N,O, and P appear to provide equally adequate solutions for our problem, the obvious suggestion for making a decision among them would be to pick first the *least* likely (as determined by the inductive leap) rather than the *most* likely. Then all except one might happen to be disconfirmed by their individual predictions, and by the process of elimination alone the remaining hypothesis would achieve a certain probability. That it would not achieve certainty lies in the fact that we can never know that we have an exhaustive list of the possible hypotheses.

As a matter of working procedure, therefore, the scientist is not quite so ready to accept the disconfirmation of a hypothesis as the logic of the situation would seem to demand. Alternatives to the complete rejection of a given hypothesis on the basis of the failure of its predictions include several possibilities which are open to us.

1) We may choose to be skeptical of the results of our testing, and simply "look again" at the facts, or perform the experiment a second time. Attention has already been called to the many possible sources of error in the acts of observation and manipulation. We often fail to observe what is there, or misread what appears to be there, or are influenced by emotional factors of which we may be at the moment quite unconscious. Furthermore, as every good scientist knows, experiments often go wrong due to carelessness in setting up the instruments, failure in the instruments themselves, impurity of the substances used, and so on. Some of my scientific colleagues have confessed that the most embarrassing experiences in their teaching careers are occasions when they have set up experiments for lecture demonstrations: having taken the greatest care to attend to every detail and having explained at length to the class exactly what will happen when they push button *x*, they find that when they push button *x*—*nothing* happens. Someone has proposed a "third law of thermodynamics" which states that if anything *can* go wrong it *will.*

2) If by repeated observations and experiments we continue to get the same negative results, we may make an alternative choice. We may re-examine the theory to see whether the predicted con-

sequences have been correctly drawn or whether, perhaps, we have made an error in deduction. A theory frequently requires the use of previously established laws and other theories in order to permit implications to be drawn.

3) If we find no error we still need not reject the theory entirely; we may modify it in some minor aspect in such a way that it no longer implies the former consequences. It is noteworthy that there are theories in science, once well established but now rejected, which retain certain aspects of their "explaining power" and are useful today under certain circumstances. Ptolemaic astronomy is still widely used today as an engineering approximation.[10] The phlogiston theory was retained as an explanatory notion long after it had been superseded by the oxygen theory.[11] The fluid theory of electricity is still useful in providing us with such terms as "current" and "flow." I have been told by biologists that often the theory of fixed species is more helpful in explaining certain living phenomena than the theory of evolution. The pragmatic factors which frequently determine the acceptance or rejection of a hypothesis, as we shall see immediately, enter at this point.

4) If no modification of the theory is possible it may be retained temporarily, as was the Newtonian theory of gravitation after the negative results of the Michelson-Morley experiment showed it to be no longer tenable. Conflicting theories, such as the wave and corpuscular theories of light, can live today for a time alongside one another, though the scientist is very unhappy about it and seeks earnestly to resolve the conflict. Only when we have failed to escape by one or more of the above devices do we accept the disconfirmation, return to the original descriptive stage, and wait for another flash of insight which will provide a substitute theory.

4. How Theories Explain

We are now ready to turn to the essential problem of this chapter: how theories explain. The problem of explaining "explana-

[10] Thomas S. Kuhn, "The Structure of Scientific Revolutions," *International Encyclopedia of Unified Science*, II, No. 2, Chap. VI, 48.

[11] *Ibid.*, p. 156.

tion" is one of the most difficult of the entire intellectual enterprise. In our first reference to this term[12] we proposed a very loose usage in order to carry us over to the point where a more careful analysis of the word could be made. We have now reached that point.

Much that has been said in the intervening pages has helped to clarify this word. Let us summarize these. (*a*) The search for explanations is the driving factor for many minds; curiosity, be it found in children or in the most profound intellectuals, will not rest with "things as they appear to be," but demands to know what they "really" are and why they are as they are. (*b*) Explanatory notions, whether they arise in flashes of illumination, or only after long periods of plodding work—collecting, describing, and idealizing—produce a feeling of profound satisfaction in the minds of those who discover them. (*c*) The most significant aspect of explanation is that there exists between that which explains and that which is to be explained an implicative relation. This has been clearly shown in the preceding pages of the present chapter, where emphasis has been placed on the fact that the logical structure of the relation between a theory or hypothesis and the predictions drawn from it (the "if—then—" relation) is precisely the same as that between the theory and the descriptive truths which were to be explained by it. To be sure, the "direction of thought" is different, since we have to get our hypothesis originally by induction before we can test it by deduction. But after the confirmation has been achieved to the desired degree, the original data, which suggested the problem in the first place, and the predicted data, which contributed to its solution, lie in the same category: all are now explained by the hypothesis.

Norman Campbell argues that while all this is true it does not get to the heart of the matter. The deductive relation is the necessary condition for explanation, but not the sufficient condition. "What else do we require? I think the best answer we can give is that in order that a theory may explain, we require it—to explain! We require that it shall add to our ideas, and that the ideas which it adds shall be acceptable."[13] This is certainly not very illuminating. But it does introduce a point which is important, and which is often referred to in discussions of the meaning of

---

[12] See above, p. 84.
[13] Norman Campbell, *What is Science?* (New York: Dover Publications, 1921), p. 83.

"explanation"—the role which "familiarity" bears in explanation. For example, one commonly hears the statement that we explain the *unknown* in terms of the *known*. This clearly makes sense, since, were there no mysteries in nature there would be no problems to be solved, and the only way to solve problems is to learn something about them which we had previously not known. But this approach to the clarification of the word "explanation" is no more helpful than the etymological one, which relates the word to *explanare* and to *explicare*, meaning in the first case "to smooth out," and in the second "to unfold" or "to reveal hidden content." The point is that in science there is a sense in which we know more about that which is explained than about that which explains it, but in another sense just the reverse is true. Salt solutions, which are well recognized by all of us, are much better known, in the sense of being directly experienced, than the intermingling of molecules by which they are explained; thus in order to discover molecules we must "smooth out" or "unfold" salt solutions. But once the molecular theory has been subjected to repeated, affirmative tests, it begins to acquire a familiarity which is greater than that of the observed facts, or, at least, different in kind. We then say that we *explain* the fact in terms of the molecular theory. *Prior* to its public confirmation the molecular theory is clear as a theory and probable as an explanation, *only in the mind of its originator.* Indeed, the enthusiasm with which a discoverer greets a novel idea is evidence both of the clarity with which he grasps its nature and of the vague anticipation of the many confirmatory acts by which he believes it will become established. *After* confirmation has been achieved this familiarity is shared by all in the scientific enterprise.[14]

For this reason a satisfactory measure of the explanatory value of a concept can be found in what are commonly called the general "tests" of a good hypothesis or theory. These are alternative ways of evaluating the techniques of confirmation, since by a good hypothesis we mean one which can successfully survive these acts, and which thereby takes on a higher and higher degree of

[14] "Remember, then, that scientific thought is the guide of action; that the truth at which it arrives is not that which we can ideally contemplate without error, but that which we may act upon without fear; and you cannot fail to see that scientific thought is not an accompaniment or condition of human progress, but human progress itself." W. K. Clifford, *Lectures and Essays* (London: Macmillan & Co., Ltd., 1886), p. 109. Reprinted by permission.

probability. No hypothesis, of course, is ever "true" or "false"; it is only confirmed or disconfirmed to a greater or lesser degree. Russell speaks of its "credibility,"[15] Kneale of its "acceptability,"[16] Braithwaite of its "reasonableness,"[17] Carnap of its "confirmation,"[18] Nagel of its "degrees of confirmation,"[19] and Ayer of the "confidence"[20] with which we may accept it. Wherever the notion of "degrees of confirmation" is introduced these quantitative variations in confirmability are generally admitted to be incapable of measurement and are therefore only crude representations of our subjective willingness to accept, at the moment, a statement for which there seems to be strong evidence.

The first requirement of a good theory is that it should have arisen out of a descriptive science and should be capable of explaining the problems which created it. This requirement will always be met when a theory occurs through the usual inductive techniques. Such a theory explains the data out of which it emerged because it is devised to do just this. Certain descriptive statements are joined, and through the inductive leap an explanatory notion is formulated from which we can deductively derive the original descriptive statements. If the theory is left in this form, of course, and no effort is made to derive other consequences from it, the explanatory power is very weak indeed. In fact the explanation may be considered to be purely verbal. Probably many attempts to explain events in terms of powers, capacities, potentialities, urges, and so on, are of this verbal character. Certainly some of the attempts at explanation by early scientists and philosophers were of this kind. Molière, in one of his plays, refers to the attempt to explain why opium produces sleep by attributing to it a "dormitive power." Verbally this accomplishes the desired result. If the statement, "Opium produces sleep," is a factual statement from which we wish to proceed inductively

---

[15] Bertrand Russell, *Human Knowledge: Its Scope and Limits* (New York: Simon & Schuster, Inc., 1948), pp. 342–343.

[16] William Kneale, *Probability and Induction* (Oxford: The Clarendon Press, 1949), pp. 250–259.

[17] R. B. Braithwaite, *Scientific Explanation* (Cambridge: Cambridge University Press, 1953), p. 120.

[18] Rudolph Carnap, "Testability and Meaning," *Philosophy of Science*, III, No. 4, October, 1936, and IV, No. 1, January, 1937.

[19] Ernest Nagel, "Probability and Degree of Confirmation," *Philosophy of Science*, ed. Danto and Morgenbesser, pp. 253–265.

[20] A. J. Ayer, *Language, Truth and Logic* (New York: Dover Publications, Inc., 1936), p. 100.

to premises, we can do so by setting up the two premises, "Opium possesses a dormitive power," and "Whatever possesses a dormitive power produces sleep"; these two latter statements will jointly imply the former, and will thus explain it. But a theory which explains only the data which suggested it in the first place is utterly useless in science. Unless we have some way of deriving *other* consequences of the existence of the dormitive power we have no way of knowing what it is. Kuhn states that certain late seventeenth-century scientists believed that the round shape of opium particles enabled them to sooth the nerves about which they moved.[21]

Braithwaite states the issue very clearly:

> A little reflexion . . . makes it clear that if the theoretical terms are defined in such a way as to make the theory logically equivalent to the facts it explains, the theory becomes merely an alternative way of stating these facts. The hypotheses . . . become translations of the empirical generalizations rather than, in any important sense, explanations of them; they do not stretch out beyond the limited number of generalizations; not only do they have exactly the same field of application, but they say exactly the same things about this field. A definition of the theoretical terms would thus sacrifice one of our principal objects in constructing a scientific theory, that of being able to extend it in the future, if way opens, to explain facts about new things by incorporating the theory in a more general theory having a wider field of application.[22]

The second requirement of a good theory, therefore, is that it should have consequences in addition to those which suggested it in the first place. This is usually expressed by saying that the theory must have *novel* consequences, we must be able through the theory to set up experimental situations which have not only never occurred in nature at any time in the past, but which might never occur apart from human ingenuity. We must be directed to areas of nature which have never been explored in order to find objects which have never before been found. The ability to foretell the future has always been an important test of true insight; some people have even attempted to prove the divine origin of the Bible by showing that it predicted many modern events. Hence we feel that a theory has met the required test if by means of it we succeed in discovering a wide variety of apparently un-

[21] Kuhn, p. 193.

[22] Braithwaite, pp. 67–68. Reprinted by permission of the publisher.

related facts, which would otherwise have completely escaped our notice, or if under the direction of the theory we are able to predict an experimental situation which is quite contrary to what we should, without the theory, have been led to expect. Anyone who is ignorant of chemistry, for example, would be quite incredulous if he were told that it made a great deal of difference whether he poured cold water in sulphuric acid or sulphuric acid in cold water; yet a knowledge of the nature of chemical substances would enable him to predict this very surprising difference. When a theory permits him to anticipate something which would be unusual and unexpected on grounds other than the theory, it receives substantial confirmation.

These two requirements might be summarized roughly by saying that a theory which explains nothing is redundant, a theory which explains a wide range of facts is of great value, and a theory which explains just what gave rise to it in the first place is *ad hoc* and tells us no more than the fact to be explained does. The ether was a hypothesis of the third kind, for when electromagnetic waves were discovered something had to be devised to serve as the subject of the verb "to undulate"; but when scientists proved to be unable to justify the existence of the ether by anything other than its wave-carrying capacity, they abandoned it as a barren hypothesis.

The third requirement of a good theory is that it be consistent with all other theories. This seems like a reasonable demand, for logic tells us that if two statements are inconsistent they cannot both be true. Though the principle is a valuable one it is difficult to apply. For example, it tells us merely that two incompatible theories cannot both be true; it does not tell us that either is true, or if one is true which one it is. Practically this causes the scientist very little difficulty since a science usually grows by small and gradual accretions, and the only theory which is commonly questioned is one which is being proposed as an addition to the large body of theories already established. Just to the extent to which a group of theories, consistent among themselves, gains in range and spread does it become increasingly unlikely that any one of the theories will turn out to be incorrect. Hence if we have a coherent body of well established theories, all knit together into a system, and if a new theory, inconsistent with this body, is a candidate for admission to the scheme, there is little doubt as to what we should do: we should usually reject the new theory in favor of the more inclusive and already confirmed system. Sometimes the new theory, in spite of its inconsistency with the rest of the system, becomes so completely confirmed that we cannot readily reject it. In such cases the confirmed theory must be accepted and the remainder of the system either rejected or more or less completely made over in accordance with the new one. This is essentially what happened in the field of physics with the advent of the theory of relativity.[23]

Another reason why the principle of consistency is difficult to apply is the comprehensiveness of the knowledge demanded of the investigator. Every theory must be shown to be consistent not only with other theories in the same science but with theories in all sciences. But no one is sufficiently familiar with the theories in all sciences to apply such a criterion. All that the scientist can do therefore is to accept the theory tentatively, but remain always alert to reject it or modify it in case it fails to fit into the broader scheme of truths.

5. Pragmatic Factors

In conclusion, we must not fail to mention the fact that hypotheses and theories are often accepted or rejected on grounds which, in a sense, fall outside man's "rationality." These are commonly called "pragmatic grounds." For example, if two conflicting hypotheses are equally successful in explaining certain facts but one of the hypotheses is simpler than the other, the former is commonly accepted. This was largely the principle on the basis of which Copernicus justified his theory of the planetary system with the sun as its center in preference to the long-held, geocentric theory of Ptolemy; each theory was equally able to explain and predict, but the system of Copernicus was very much simpler[24] mathematically than that of Ptolemy. Furthermore, as has

[23] Kuhn, especially Chaps. VI, VII, and VIII.

[24] For a further discussion of *simplicity* as a principle for accepting scientific theories see: Mario Bunge, *The Myth of Simplicity* (New York: Prentice-Hall, Inc., 1963); "The Role of Simplicity in Explanation," *Current Issues in the Philosophy of Science*, ed. H. Feigl and G. Maxwell (New York: Holt, Rinehart & Winston, Inc., 1961), pp. 265–285; and R. Harré, *An Introduction to the Logic of the Sciences* (New York: The Macmillan Company, 1960), Chap. 7.

been suggested,[25] early theories which have been quite completely disconfirmed, are still employed occasionally as explanatory devices, perhaps useful in teaching; pragmatic factors enter into such decisions. Sometimes one theory may be preferred to another simply because it is notationally more convenient. Pragmatic considerations determine the precise moment when a certain scientist will say to himself that a given theory has been sufficiently confirmed to be allowed to enter into the body of established science. Some scientists wish to "play safe" and withhold commitment until the confirmation has been so firmly established that an objector would be judged by his colleagues to be a pathological case; others are more venturesome and readily accept new theories even though the evidence in favor of them is weak and inconclusive. Sometimes a practical problem arises in which immediate and decisive action is called for, even though evidence in support of alternative modes of procedure might be equally balanced. A simple example is the doctor who is confronted by a patient with a serious case of what may or may not be appendicitis; the doctor must decide whether to operate immediately in the hope of saving the man's life by removing the appendix but with the risk attendant on all surgery, or postpone such action until more tests can be made but with the chance that the man cannot survive the delay. Here, again, pragmatic factors would produce the decision, whether it be the right one or the wrong one.

6. Kinds of Explanation

Once the problem of the explanatory character of theories has been understood, the question as to what *kinds* of explanation are to be found in science is easily answered. Wherever a theoretical statement implies a descriptive truth, and wherever, in the words of Campbell, quoted above, the theoretical statement "adds to our ideas" and "adds ideas which are acceptable," there explanation exists. There are, however, many commonly employed modes of explanation. In the ideographic sciences, for example, explanation is largely *of* objects and events at particular times and places *by* other objects and events at other times and places. Here causal laws and functional and statistical correlations provide the chief

[25] See above, p. 125.

explanatory *aids*, and the language of such science abounds in proper names, dates, clock times, and latitude and longitude indications, as well as characterizing symbols describing what is to be found in these times and places. In the nomothetic sciences, explanation of individual objects in terms of other individual objects is less prominent. Instead, descriptive concepts are explained

FIGURE 9

EXPLANATORY SCIENCE[26]

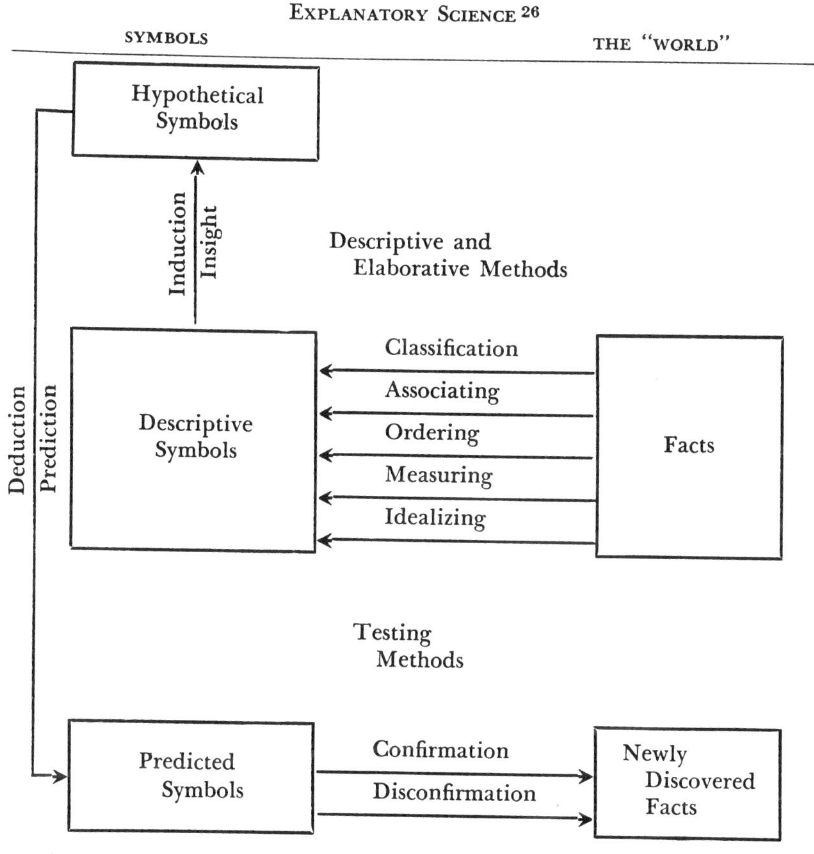

[26] The use of diagrams in the portrayal of the structure of the scientific method is a common practice. Some illustrations are the following: M. Bunge, *Metascientific Queries* (Springfield, Ill.: Charles C. Thomas, Publisher, 1951), pp. 85, 94; J. G. Kemeny, *A Philosopher Looks at Science* (Princeton: D. Van Nostrand Company, Inc., 1959), pp. 86, 168; Henry Margenau, *Nature of Physical Reality* (New York: McGraw-Hill Book Company, Inc., 1950), pp. 85, 106; and F. S. C. Northrup, *The Logic of the Sciences and the Humanities* (New York: The Macmillan Company, 1947), p. 145. All such diagrams are icons, which attempt to represent logical structure by means of spatial relations. As such their limitations should be clearly recognized: they must be highly abstract and schematic; they should be sufficiently general to include the ex-

by theoretical concepts, repeated correlations of objects and events are explained by general laws, and general laws are explained by higher laws. Where causal explanations are found they tend to be either "mechanistic" or "teleological"—the former primarily in the physical sciences but also in the biological sciences, and the latter primarily in the biological sciences but dominantly in the behavioral sciences. In all sciences, both ideographic and nomothetic, there are explanations of wholes by their parts and of parts by their associations in larger wholes which contain them. In all sciences where quantitative explanations are available their use is preferred to qualitative explanations.

The conclusions of this chapter can now be summarized in a diagram. We have shown that a group of descriptive symbols (temporarily called a "descriptive science") becomes transformed into an explanatory science when (*a*) we add to the descriptive statements other statements, not known to be descriptive in character, but about hypothetical entities derived from the described entities by the method of induction; and when (*b*) we confirm these statements about hypothetical events and objects by showing that other statements about presumed facts can be deduced from the statements about the hypothetical events and objects, and that the statements about the presumed facts can be confirmed by the observed existence of these facts. An explanatory science, therefore, includes the descriptive science out of which it arose, and also statements about theoretical entities, statements about predicted facts, and either the facts so predicted or alternative facts which make their existence impossible.

---

perimental and the non-experimental sciences, the quantitative and the non-quantitative, and the ideographic and the nomothetic; and they should make clear that their static form does not prevent the portrayal of the "free flowing" movement of science as it actually occurs in the behavior of the scientist. Figure 9, of course, represents only explanatory science, and must be taken with Figure 7, p. 80, Figure 8, p. 109, and with Figure 10 to follow on p. 147. My use of arrows to represent the "movements" of science is designed to correspond to the order which I have chosen in the book to discuss the "elements" of science as they constitute a "structure." It should not under any circumstances be taken as a representation of a view that science grows through a fixed temporal series consisting of observation, description, hypothesis-formation, and testing. It attempts to show the *logical* structure of science, not the *temporal* order of the scientific pursuit.

7

Formal Science

Had this book been written a hundred and fifty years ago the topic of the present chapter might not have been included. While the main ideas here to be developed go back at least as far as Euclid, the knowledge of the nature and structure of deductive systems is a comparatively recent acquisition. The topic is inserted in our discussion for the double purpose of throwing greater light on the distinction between descriptive and explanatory science, and of giving adequate recognition to mathematics in the scheme of the sciences.

We have had occasion to point out repeatedly that descriptive science represents knowledge in a somewhat unsatisfying and unstable form. While accurate description provides important information which is absolutely essential to the development of a complete science, it does not seem to represent the final goal which we are striving to reach in our study of nature. Unless we deliberately restrain ourselves we discover that our minds *will* venture out into the realm of imagination in the attempt to explain and account for the facts which we have collected and described. A descriptive science will not remain purely descriptive. We want more facts, we like to speculate, and we discover even in descriptive science those rudiments of logical organization which can be so readily expanded into an embryonic theoretical science.

But if description is not the final stage in science, perhaps explanation is. However, even this seems not to be the case. Explanatory science, as we hope to show in this chapter, itself tends to change into something else. The desire to explain, which originated at the descriptive level, will not be satisfied with anything short of final and complete "explanation"—something which even explanatory science is not prepared to offer.

1. Explanation and Implication

In the attempt to explain further what explanation is let us suppose that it exhibits two aspects—one logical and the other psychological. The logical aspect is the implicative relation holding between that which explains and that of which it is an explanation. The psychological aspect is the total complex of feelings of confidence, belief, degree of conviction, familiarity, explicit and implicit knowledge, novelty in explanation, and a host of other terms expressing mental attitudes toward statements or toward the facts which they presumably assert. We suggested in section 4 of Chapter 5 that the incentive to expand descriptive science might have arisen in the discovery that even *within* such a science some of the truths are actually related by implication. In the questionnaire we learned that the statements A,B, and C were all true, and then by further inspection we learned that A and C, taken together, implied B. This led to the hope that implicative relations could be discovered to hold throughout nature and human behavior.

In the partial fulfillment of this hope we succeeded in constructing explanatory sciences by expanding descriptive sciences through the invention of hypothetical and theoretical notions. As more and more of such notions were introduced, as they were given wider and wider generality, and as more facts were discovered through their use, we found that explanation played a more and more important role in science. The system took on a greater and greater comprehensiveness. But it fell far short of perfection in this regard. There were loose ends, brute facts, statements which were descriptively true but could not be fitted into the scheme. In some cases we succeeded in integrating a small body of facts in terms of a theory which was proper to them. But this remained an island of order in a sea of disorder. Frequently separate bodies of facts could be unified, each in terms of its proper theory, but the groups of facts were not interrelated in any significant way and the theories had no principle of unification.

All these inadequacies—which are readily recognized in any explanatory science—lead us to speculate concerning an ideal state of science in which the logical organization is perfect and complete. This would presumably be a science in which all descriptive statements would be capable of being derived by pure deduction from a single theory, this theory itself being so constituted as to be as simple as possible and to involve no internal contradictions. Such an ideal would be achieved in physics, for example, if from a theory of the electron or of some other elementary particle, we could deduce all the facts of physics—dynamical, acoustical, electrical, optical—and perhaps even the facts of chemistry as well. Such an ideal would be achieved for the totality of the sciences if we could discover a single, unifying theory from which all the facts of all the sciences could be inferred.

If we now refer to our earlier discussion,[1] where the notion of "implication" was subjected to preliminary analysis, we shall recall that when P implies Q, P bears to Q a relation which Q does not, unless further information is provided, bear to P. There is a directional feature of the relation which we express by saying that it is "non-symmetrical." (This does not mean that the relation is asymmetrical, because, as we stated in our definition of an ordering relation,[2] a relation is asymmetrical only when if x has the relation R to y then y *cannot* have the relation R to x. A relation is non-symmetrical when if x has the relation R to y then y may or may not have the relation R to x.) We found that this required us to admit that if P is true Q must be true, but if Q is true P may or may not be true. (It also requires us to admit that if Q is false P must be false, but this is of no significance in the inductive movement of explanatory science since we never start with false descriptions and try to explain them in terms of

[1] See above, pp. 90–95.

[2] See above, p. 67.

hypotheses.) The relation of implication is called "formal" because when P implies Q this relationship is completely independent of our *knowledge* of the truth or falsity of either P or Q. By formalizing the relation we isolate its "logical aspects," and permit it to be considered entirely apart from what we called a moment ago its "psychological aspects"—whether or not P or Q should be "believed," whether one or the other is more "familiar" to us, whether or not anything "novel" emerges as we pass from one to the other, whether either one explicitly or implicitly "contains" the other, and so on. When we drop these considerations, implication becomes a bare structure or form, itself without content but capable of taking on content when we insert P's and Q's having associations with a world which would make them true or false, and with our mental attitudes which would make us believe them, doubt them, learn from them, and explain by means of them. Our science becomes "empty" of content, and breaks its contact with the world of objects and events.

Furthermore, in our development of a more perfect explanatory science we found another factor working toward the separation of our symbolic scheme from the world. Descriptive processes, such as classifying, associating, and ordering, seemed only to demand more elaborate operations, such as measuring and idealizing. Both of these—particularly the latter—involved us in more complicated and elaborate activities which resulted in symbols almost completely lacking in direct, descriptive reference. At the level of the measurement of simple qualities, such as time, space, and mass, the process seemed direct enough, and we therefore felt entitled to say that the numbers did in some sense represent the qualities which had been measured; but if we asked by how many units our enjoyment of Beethoven exceeded our enjoyment of Bach, we saw how artificial the operation of measurement could be. Measurement is truly descriptive just to the extent to which we can specify clearly the processes by which numbers are attached to objects and events, and to the extent to which these numbers can be considered to be *meaningfully* attached to the objects. In the case of idealizations, however, descriptive value seemed to have been more or less completely sacrificed. All we could say was that the world "approximated" to the ideals of our descriptive scheme; there are no perfect gases but there are relatively perfect gases which, within certain limits of variation, do obey the gas law;

there are no individuals who have been completely isolated from social contacts, but there are individuals who have lived as hermits at least part of their lives and what we say about perfectly isolated individuals applies to them to a certain degree of approximation.

The existence of such processes as measurement and idealization, together with the recognition that the implicative relation between statements does not depend on the actual truths of the statements themselves, lead us to look for other processes which indicate a break between the facts of a science and the symbolic system which attempts to portray these facts. A simple inspection of the science of algebra shows one such process. The method involved is abstraction. This is the technique used in the process of simple classification, which is one of the most basic of the descriptive methods. Here, however, we are concerned not with the mere grouping of objects into classes on the basis of directly observed similarities but with a process of *high* abstraction which involves a more or less complete neglect of the specific differences of objects and an emphasis on their most general properties. Let us look, first, at arithmetic. A fairly high degree of generalization is involved even here—which may explain why many of us as children had some difficulty grasping the subject. When we first learned that $2 + 2 = 4$ we were taught to think concretely—in terms of books, stones, tables, and the like. But we soon learned that *what* we are talking about in arithmetic really doesn't matter; *any* two things plus two things equals four things. This important step is taken by an act of abstract generalization; the number "2" is no longer restricted to a pair of shoes, a married couple, ham and eggs, black and white, or any other specified pair, but is recognized as applying to something which is common to all these classes. The step involved here is not an easy one and represents, so far as the race is concerned, a highly significant advance over our primitive ancestors. But the step to algebra is still more difficult. At this level we must grasp the meaning of something which is common to 2,5,4,3,7,6, and all the numbers, even including the fractions, the negatives, the irrationals, and the imaginaries. We were told by our teachers that x represents *any* number. For most of us x was plainly not a number at all but a letter, as any reasonable person could see. We were told, however, that x is used when we want to talk about a number without bothering to say which number it is. Bertrand Russell, whose great skill in abstract thinking is well known, confessed to me once in a personal conversation that when as a youth he was first presented by his tutor with the notion that x could represent *any* number, he was very much puzzled. In fact he thought for a long time that his tutor really knew what number x was but for some strange reason wouldn't tell him. To be able to think without thinking about anything in particular is to achieve a high degree of skill in abstraction—to think, in other words, about the form of things rather than their content. We seem to have to travel such a great distance in going from a pair of shoes to x that it seems rather absurd to say that the letter x in any sense "describes" the object which originated it.

Such considerations as these suggest that the descriptive ideal certainly does not dominate science. Not only in the admission of measurement, idealization, and generalization into science but in the use of theories and hypotheses has the descriptive ideal been forsaken. The essential point about numbers, idealizations, and abstractions is that they are not known to exist in the usual sense.[3] But this does not seem to prevent us from talking about them. There seems to be involved in all explanatory science a realization that descriptive accuracy is not, after all, the most important thing. To be sure, we want our statements to describe actual objects—if this is at all possible. But once we have recognized the power of deductive organization, once we have seen what an explanatory science can do in the way of rendering the world intelligible and enabling us to predict new facts, we realize that close correspondence with the world of fact is not really so important as it had seemed to be. We even recognize that where descriptive accuracy and effective deductive organization cannot both be achieved the obvious thing to do is to sacrifice descriptive accuracy.

2. Abstract Character of Formal Science

This abandonment of the descriptive ideal of science can be beautifully illustrated by a stage in the history of geometry.[4] Prior to

[3] This question will come up for further consideration on pp. 145–147 and 154.

[4] A readily understandable exposition of the nature of non-Euclidean geometries can be found in C. G. Hempel, "Geometry and Empirical Science," *Readings in*

the early part of the last century no one had felt that there were any reasonable grounds for questioning the descriptive adequacy of this science. It had arisen, as the etymology of the word "geometry" indicates, out of the attempts of man to measure land and possibly to determine property rights, as has been suggested, after the annual inundations by the Nile had obliterated previous markings. Pythagoras, Euclid, and their successors had refined the science, made it more exact, and given it the rudiments of logical structure. The fact that it was concerned with idealizations, such as point, line, and circle, all of which were obtained from the objects of observed experience by operations of refinement, was well recognized. If a point had mere position but no dimensions it could not be seen, for if we could see such a thing it would necessarily have some size. Demonstrations which were made with the use of figures drawn on paper or on the blackboard were known not to apply really to these figures, but to their idealized counterparts, existing only in the imagination. There was no question, in other words, of the applicability of the notions of geometry, if only by approximation, to the world in which we live. They were of unquestioned use in building bridges, measuring cloth, constructing railroads, and erecting houses. The study was generally recognized to be, therefore, a splendid example of an explanatory science—that is, a descriptive science which through the use of idealization and measurement had to a certain extent lost its descriptive value, and through the use of insight and discovery had taken on a rough deductive structure.

During the nineteenth century geometry experienced some severe shocks. Two geometricians, one a Russian named Lobachewski, and the other a German named Riemann, made proposals which required a fundamental change in our attitude toward this science as a description of the world. Among the basic ideas in the Euclidean system of geometry was a statement which has been shown to be equivalent to what is now called the "parallel postulate." This asserts that from a point outside a line only one line can be drawn parallel to the given line. Such a statement seemed to Euclid to be self-evidently true. Even if it were not, however, it was required in proving some of the theorems of the system and thus was demanded in the same way that a hypoth-

---

Philosophical Analysis, ed. H. Feigl and W. Sellars (New York: Appleton-Century-Crofts, Inc., 1949), pp. 238–249.

esis is demanded in any explanatory science. Lobachewski and Riemann conceived the possibility of denying this statement and replacing it by another. Lobachewski chose the statement that through a point outside a line *an infinite number of lines* can be drawn parallel to the given line. Riemann showed that still another statement could be selected: "Through a point outside a line *no* line can be drawn parallel to the given line." Each geometrician then proceeded to construct a geometry by combining his special postulate with the others of Euclid's geometry. The surprising result was that both of them succeeded in constructing such systems, and that each system proved to be internally just as consistent as the Euclidean system. Each had its group of underived notions, from which, by pure deduction, all the theorems of the system could be proved. On grounds of logic, therefore, there was no way of deciding between the alternative schemes. But they were inconsistent with one another, not merely in the statement of the parallel postulate but in many of the theorems which followed from the postulates. For example, as the reader will recall from his high school geometry, in the Euclidean system the sum of the angles of a triangle is equal to 180°. But in the Lobachewskian system it is always less than 180°, and in the Riemannian system the sum is always greater than 180°. There are other differences which need not be noted at this point. The question which immediately arose, of course, was this: Which system of geometry is true of our world? They cannot all be true, for the sum of the angles of a triangle cannot be at once equal to, less than, and greater than 180°. Measurement of actual triangles cannot help to decide the issue, for only very rarely do we find the sum equal to 180°; our measuring instruments are so inaccurate that we usually find our results to be sometimes greater and sometimes less, with no possibility of deciding which is the true value. The same is true of the postulates. Suppose we attempt by actual construction to determine how many parallels can be drawn to a given line from a point outside; surely we should be able to decide whether it is one, two, or none. The difficulty is, however, that we cannot draw *any* lines, for these are purely ideal entities which cannot be produced with chalk or ink.

As a result of these studies the conception of mathematics[5] as

---

[5] In recent years much literature has accumulated on the nature of mathematics, and of formal sciences in general. Some references, of varying degrees of difficulty, are

a science was changed in two fundamental ways. In the first place, it became apparent that geometry (and, in fact, the other mathematical sciences) does not talk about the world at all—or, at least, talks about it in such general terms as to make the application in concrete cases extremely hazardous. Geometry is a purely deductive scheme, having all the delightful logical properties of such systems, but not necessarily specifically applicable to the world in which we live. The connection with particular facts in such a science has been broken completely, and it has ceased to be descriptive; it talks about none of the objects of our ordinary experience and is concerned only with a dream world or even with no world at all. The possibility of such a science as this shows us that the human mind has the capacity for taking at random any group of ideas, actual or fanciful, combining them, and deriving their consequences by pure reasoning and thus constructing a system of underived and derived notions which is internally consistent, which obeys the rules of logic throughout, but which has nothing specific to say about the world. Of such a system of notions we are able to say only that if we admit the basic ideas we must, if we are to be consistent in our thinking, admit also the derived ideas. But we are not compelled to accept the basic ideas, for they may be definitely contradicted by our experience, or they may be so abstract and general in character that we are unable to tell whether they apply to our world or not.

In the second place, the view can no longer be held that the basic truths of mathematics (the axioms and postulates) are in any sense obvious or self-evident; for the three alternative formulations of the parallel postulate cannot *all* be self-evident if they are incompatible with one another. The notion that truth can often be determined by a kind of intuition, or direct inspection, has played an important role in the history of thought, and many philosophical systems have been erected on such foundations. But its unreliability is obvious for many reasons. A statement may be self-evident to one person and not to another, and when a conflict of this kind arises there is literally *no* method of resolving it.

the following: *Ibid.*, Sec. III; *Readings in Philosophy of Science*, ed. H. Feigl and M. Brodbeck (New York: Appleton-Century-Crofts, Inc., 1953), Sec. II; R. L. Wilder, *Introduction to the Foundations of Mathematics* (New York: John Wiley & Sons, Inc., 1952); F. Waismann, *Introduction to Mathematical Thinking* (New York: Frederick Ungar Publishing Company, 1951); Bertrand Russell, *Introduction to Mathematical Philosophy* (New York: The Macmillan Company, 1919).

Many beliefs — that the earth must be flat because otherwise people would fall off; that everything that happens must have a cause; that the world would continue to exist even if no one were perceiving it; that every statement must be either true or false — have been shown to be, if not false then at least debatable, and hence not self-evident. A story is told about a mathematics teacher (presumably of the old school) who put a formula on the blackboard with the announcement, "This statement is self-evident." Immediately a student raised his hand and asked, "*Is* that statement self-evident?" The instructor hesitated, looked at the formula, and then announced, "Class dismissed." He then went home, thought about the statement for the entire weekend, and when the class reassembled on Monday proudly announced, "Yes, the statement *is* self-evident." How can a statement be self-evident if one has to think about it for three days in order to detect that fact? Truths which seem to be self-evident often have the same kind of certainty about them that new hypotheses do when they arise through the flash of insight; and we have seen that their origin in this manner is no guaranty of their truth.

Other examples of such abstract systems can be found in rather surprising places. Consider a game, such as chess. The underived ideas are the men, the board with its arrangement of squares, and the rules according to which the men are placed on the squares, moved about, and taken. The playing of the game consists in deriving, from these assumed notions, a certain position of men in which the king of one player is unable to avoid capture by the pieces of his opponent. This is equivalent to one of the derived notions of a deductive scheme, and the problem is to obtain it from the original conditions by rational use of the rules of the game. The logical structure is clearly evident. But what of the descriptive character? There is certainly something of this in the naming of the pieces, and in the vague resemblance of the playing of a game to the waging of a battle. (That the queen is much stronger than the king in chess may or may not impress the reader with the descriptive accuracy of the game.) But in playing the game there is no thought of the representative character of the pieces any more than there is in the playing of bridge where kings, queens, and jacks are involved. Neither chess nor bridge is "about" anything, and is therefore in no sense descriptive; it is simply a pastime in which opponents starting with the same conditions, match wits with one another to see which can be the first to derive a certain desired conclusion.

Even a highly fanciful story, such as *Alice in Wonderland,* is nothing but a somewhat imperfect example of a deductive scheme having no descriptive value. To be sure, we have to operate with an initial condition — falling down a rabbit hole — which seems to have no very significant logical consequences. But we should not forget that we have another condition — a highly imaginative girl — and, what is more important, we are required to use "rabbit hole logic" in making our deductions. Under these circumstances what happens to Alice is just about what one would expect. Much fiction is of this kind. In any such book there is an initial statement of conditions, a problem posed, and a resulting statement of how these conditions operate in the life and action of an individual or group. Even in non-fiction books, books in which there is exposition or argument, there are the elements of deductive structure. One frequently comes upon statements such as "This is self-evident," or "This is obvious," and then one often finds that other statements are proved in terms of these. But the deductive structure is very confused and obscure, and many writers commit the unforgivable sin of proving A in terms of B, and then, at some later point, proving B in terms of A. Such circular reasoning is never possible when the logical structure is clearly evident. One very interesting example of a document having the beginnings of a logical structure is the Declaration of Independence. Early in this document, it will be recalled, there occurs the statement, "We hold these truths to be self-evident, that all men are created equal, that they are endowed by their Creator with certain inalienable Rights, that among these are Life, Liberty, and the pursuit of Happiness." The document then goes on to point out that when such rights are violated by the established government, revolt and the creation of a more just one are the duty of the oppressed peoples. The logical structure is clearly evident. To be sure, in view of the conditions existing at the time we may assume that the document had also descriptive reference. It is therefore not a pure deductive science but an example of a set of symbols which have achieved a certain logical structure without sacrificing descriptive adequacy.

When a science has attained deductive structure at the sacrifice of descriptive adequacy it is called, variously, "rational," because its organizing principle is reason; "formal," because it has no specific content; or "autonomous,"[6] because it exists in its own right and does not depend on particular facts. Mathematics as a purely uninterpreted deductive scheme is identified with formal logic, and we shall henceforth assume this. The logical structure exhibited by a mathematical scheme is indicated by the presence of such words as *implies, if . . . then . . . , both, or, not.* If one asks what there is "in the world" corresponding to these logical terms, he is proposing a problem which is indeed puzzling. One cannot say that they refer to nothing at all, for mathematical and logical systems can often be *applied* to the world, and this would seem to indicate a structure to which they can be attached. But to say that this logical structure "describes" the particular facts of the world seems quite inappropriate. Alternatively one might argue that the "total world" is divided into two parts—a spatio-temporal realm consisting of particular objects and happenings, and a non-temporal, non-spatial part (sometimes called a world of "subsistence," in opposition to the world of "existence") which is constituted by abstract structures.[7] Fortunately the solution of this troublesome problem is not necessary for our task. We shall therefore say that mathematics and logic are "about the world," not in the sense in which descriptive and explanatory sciences are, but in a unique sense which we can formulate by saying that they *refer* to *abstract structures.*

The structure of a formal science is as follows: It consists of a body of underived ideas and a body of derived ideas. The derived ideas are the theorems of the system; the underived ideas are the undefined terms, and the axioms or postulates. If the goal of such a science is the complete proof of all the theorems in terms of the axioms or postulates it is clear that these latter notions must be taken as incapable of proof; we cannot prove A by reducing it to B, then prove B by reducing it to C, and continuing this process forever. We must always start with something which is taken as true, either because it is self-evident, because its opposite is inconceivable, or because we have simply for the moment *decided* to

---

[6] C. J. Keyser, *Thinking About Thinking* (New York: E. P. Dutton & Co., Inc., 1926).

[7] An interesting recent discussion of the problem of what mathematical symbols symbolize is to be found in *Philosophy of Science: The Delaware Seminar,* ed. B. Baumrin (New York: John Wiley & Sons, Inc., 1963), Part III, "Philosophical Aspects of the Foundations of Mathematics."

use symbols in such a way as to *make* it true. Similarly if the goal of such a science is to explain all the complex ideas in the theorems by means of the simple ideas in the axioms or postulates, these ideas must themselves be taken as indefinable; we cannot define A in terms of B, then define B in terms of C, and continue this process forever. We must start with ideas whose meaning is initially clear, or with symbols which mean nothing at all, and then proceed by means of the postulates to make statements about them and thus gradually give them content.

Such a science is represented, according to our general plan, in Figure 10.

FIGURE 10
FORMAL SCIENCE

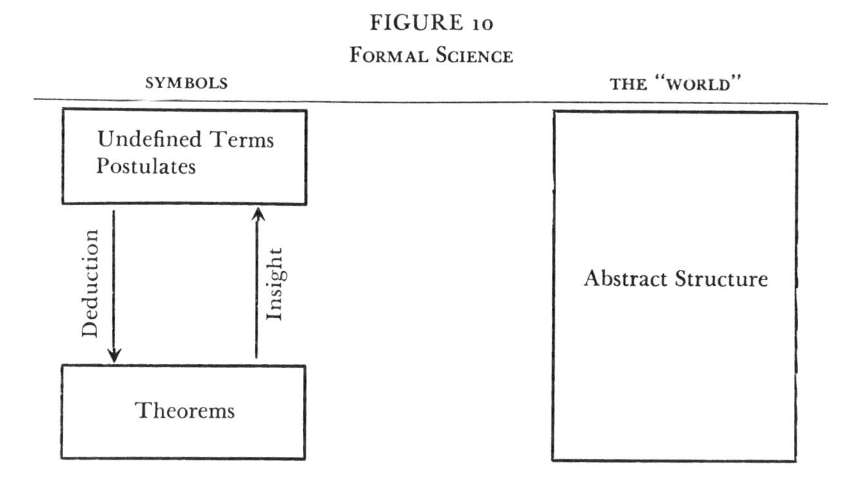

3. Example of Abstract Science

For purposes of illustration an example of a simple formal science is given below:

I. Underived notions
  A. Undefined ideas (assumed to be understood)
     elements: $x_1, x_2, x_3. \dots$
     relation: R
  B. Postulates (assumed to be true)
     1. There is an x such that no other x bears the relation R to it.
     2. No x has more than one x bearing the relation R to it.
     3. Each x bears the relation R to just one other x.
     4. No x bears the relation R to itself.

II. Derived notions (theorems)

1. There are infinitely many x's.

Proof: Assume that there is only a finite number of x's, say 6. Postulate 1 tells us that there is a "first" element, i.e., there is no x which has the relation R to this "first" x. From postulate 3 we know that there will be a "second" x, a "third" x, until the class is exhausted. Now we know by postulate 3 that the "sixth" x must bear the relation R to some x. By postulate 4 we know that this x cannot be the "sixth" x, so it must be another x. We have already shown that this other x cannot be the "first" x, and by postulate 2 we know that it cannot be the "second," nor the "third" . . . to the "sixth," because each of these already has an x which bears the relation R to it. Hence there is no x to which our "sixth" x can have the relation R, and postulate 3 cannot be satisfied. This shows that we made a mistake in assuming that there is only a finite number of x's; instead there must be an infinite number. Hence our theorem is proved.

Note that there is no specific interpretation of this rational science. The notions of "element" and "relation" are purposely left undefined. Such a science can be made descriptive by giving a concrete interpretation to the notions contained in it. Let us suppose that the x's are *boys*, and R is the relation *tags*. Then our science could represent a game called "tagging" which a group of boys might decide to play. Each boy would then become a "tagger" and would be required to tag only one boy. This boy must, in each case, be one who has not already been tagged by another boy. No boy could tag himself. One boy, called the "chief tagger," must tag as the other boys do, but cannot himself be tagged by any boy. Our boys would then find, probably to their sorrow, that they could never play this game unless they had an unlimited group of "extras" who would have to be called into the game one by one in order to make the rules apply.

Other interpretations of this abstract system would be possible. We might take a group of dogs and apply the principle *dog eat dog*. We might arrange books on a shelf by the relation *stands immediately to the left of*. We might consider the x's to be children of certain parents with the relation *next older than*. (There would then have to be an infinite number of children!) If no interpretation is made the science is ordinarily said to be completely abstract and without content. This is not quite true, since, as we have already seen, the notions of "element" and "relation" have a certain reference to abstract structures. But because of their high generality they have no concrete reference and the science is usually, therefore, considered to be a purely formal science.

4. Continuity of the Three Forms of Science

We are now ready to summarize the results of our study up to this point. With the addition of formal sciences to our list we are in a position to contrast and compare the three basic types of science. Rutherford is reported to have said there are only two sciences—physics and stamp-collecting. In view of the development of the formal sciences since the days of Rutherford we shall now be entitled to say that there are only three sciences—physics, stamp-collecting, and checkers. Our summarizing remarks may prove to be partly repetitious of earlier conclusions, but they should enable us to see the types of science in their true perspectives and in their proper relations to one another. Such an undertaking seems essential if we are to evaluate the role of science in living.

The point of view which we have taken has emphasized the continuity of the three forms of science.[8] They can be readily arranged in a three-termed series, with descriptive science at one end, formal science at the other, and explanatory science, since it shares features with each of the extremes, occupying the middle position. Not only is the arrangement continuous but there is a rough temporal development from descriptive to formal science. This does not, as we have seen, correspond closely with the actual

---

[8] The idea of a continuity in the three forms of science was first suggested to me, though in a very nebulous form, many years ago when I became acquainted with the writings of Émile Boutroux, F. L. Ravaisson, Jules Lachelier, and Jacques Rueff. Although these French thinkers are not generally read today, and are scarcely even mentioned in histories of philosophy, they seem to me to have stated, implicitly rather than explicitly, the essential problem of the scientific understanding of the world, viz., how to reconcile the necessity of mathematics with the apparent brute contingency of the actual world in which we find ourselves. I believe that this reconciliation has been achieved, to the extent to which it is possible, by explanatory science.

historical development, though in terms of the broad progress of science through the centuries a surprisingly close correlation exists. If the parallel between the temporal and the logical were perfect, all early science would be descriptive and all modern science would be formal. The existence of highly developed formal sciences among the Greeks, and of significant descriptive sciences today sufficiently refutes this claim. Fact-finding, search for explanations, and detection of logical structure go hand in hand. All we have attempted to show is that there is a strong tendency for a descriptive science to grow into explanatory science, and for this to transform itself into formal science.

Whether formal science represents the perfection of science, or whether the loss of descriptive value at this level tends to force science to retreat to the stage of explanation, are matters which are not directly relevant to our discussion. Many would insist that the transformation of a descriptive into a formal science is a progressive movement—a movement involving the improvement or betterment of the science. Others, the traditional positivists,[9] would deny this. In fact, at least three value attitudes, all determined basically by matters of temperament, can be found as principles for the interpretation of this transition. Those who feel at home in the presence of the concrete objects of this workaday world and who are uncomfortable in the realm of logical abstractions will prefer descriptive to formal science, will detect the germs of decay in descriptive science when the first attempt is made to introduce hypotheses and theories. Those at the other extreme, who find delight in the rare atmosphere of high abstractions and are confused by this higgledy-piggledy, job-lot of a world see both descriptive and explanatory science as preparatory to the development of formal science, which can be the only satisfactory study of the world. Still others, endeavoring to take a middle-of-the-road attitude, appreciate fully the logical inadequacies of descriptive science and judge the transition to explanatory science to be an important movement in the right direction; but they condemn further refinement of logical systems as the mere playing of games—as a kind of mental gymnastics. They see any movement from explanatory to formal science as definitely regressive, and therefore to be strongly resisted. For the first group, descriptive science is science in its purity; decay and pollu-

---

[9] See above, p. 85.

tion begin at the first appearance of the desire to explain; formal science is really not science at all. For the second group, descriptive science is embryonic; explanatory science represents birth and growth; formal science is ripe maturity. For the third group, description is science preparing to become explanatory; explanatory science alone is true science; and formal science is explanatory science which has overshot its mark.

The contrasting viewpoints of the three types of science can be summarized and clarified by examining the manner in which each solves certain problems which are common to them. Two such problems may be selected. First, how is the meaning of words and other symbols clarified? Second, how is truth established? All of the sciences—descriptive, explanatory, or formal—are bodies of symbols. But symbols, in order to be symbols and not just marks, must have meanings, and combinations of symbols, when they are statements, must be not only meaningful but true or false. If we can understand how each of the three types of science determines, on the one hand, meaning, and, on the other, truth, we shall be in a position to see precisely how the sciences differ from one another.

The fundamental method for clarifying symbols in descriptive science is *pointing*. This is, of course, also the technique of common sense. If we wish to explain what is meant by the word "airplane" to an individual who has never seen one we might resort to definition, but we should not feel that we had really cleared up the word until we could point to an airplane and say, "Look; that is what I mean." We all know how difficult it is to describe the taste of a certain food to a person who has never eaten it. But if we can give him the food and let him have the experience of tasting it we shall be successful. The method here used is a kind of indirect pointing. We commonly recognize how hard it is to make a blind person understand what we mean by color, or a deaf person what we mean by sound; the difficulty resides in the impossibility of using the method of pointing. In the last analysis all symbols, except such syntactical and logical words as *the, all, is, implies, and,* and *not,* must get their meaning, directly or indirectly, either from objects in the world or from "felt" experiences of various kinds. Gestures are commonly used to indicate the approximate areas within which the object we are talking about is to be found. This direct reference to experienced objects is the essential method of descriptive science. To be sure, because of the restricted way in which the method can be used, certain less direct techniques are employed. But all these—observation, interpreting reports, experimentation, classification, association, ordering, and even measurement—are, in essence, applications of the method of pointing. When the objects are not present we wait for them; when they will never occur under natural conditions we manufacture them. When they are present we describe them and measure them. These are all techniques for confronting ourselves with objects so that we may point to them when the occasion arises.

In formal science the method of clarification is quite different. Pointing is never used, for this directs attention outside the system of symbols and a purely formal science has no "outside," except in the special sense indicated above.[10] Meaning and clarification are merely matters of the interrelationship of symbols. The most common of these is *definition*. Definition takes many forms in deductive systems.[11] It is always, however, a process of relating the meanings of complex symbols to the meanings of simple and more elemental symbols. Such a process is successful only if the elementary notions are themselves more clearly understood. How, then, are *these* clarified? Usually they are not clarified at all. They are accepted as indefinable and irreducible terms, incapable of reduction to anything more elementary. Sometimes certain of the basic ideas are taken over from other fields of study, as in the case of the mathematician who employs ideas drawn from logic, or of the physicist who uses ideas drawn from mathematics. The formal character of the science in which such symbols occur assures us that we should not be too much concerned with specific meaning. Certainly if we are trying to clarify a highly general notion, such as "element," or "class," or "relation," we should be misled if we tried to locate the meaning in a specific and concrete object. For example, if we attempt to clarify the meaning of "class" by pointing to a class of men, we can have no assurance that our hearer will attend to that which is common to all classes —which is really what we have in mind—rather than confine his reference to the actual physical group which is present, or to

---

[10] See above, p. 145.

[11] For example, Irving Copi, *Symbolic Logic* (New York: The Macmillan Company, 1954), pp. 155, 167–169, 191.

the specific characteristics of human beings. In general, then, the method for the clarification of symbols in formal science is one of exploring the interrelations of the symbols within the symbolic scheme. Symbols are defined by other symbols, not by things.

Since explanatory science is to a certain extent a combination of descriptive and formal science, the techniques for the clarification of meanings are both *pointing* and *reduction of hypothetical notions to descriptive symbols.* But because explanatory science has departed somewhat from its factual foundation, pointing techniques are not entirely adequate; and because it has not yet achieved perfect deductive structure, reduction techniques fall somewhat short of complete success. For example, to the extent to which idealizations are an important part of explanatory science the method of pointing does not work. We cannot point to a perfect lever or to an ideal gas. All that we can do is to point to approximate exemplifications of such notions and suggest that what we are really talking about is this sort of thing after it has been subjected to certain refinements—just like this only better. Hypothetical and theoretical notions are even more difficult to define; we cannot use pointing, for the objects to which they refer are usually unobservable. Such symbols are meaningful to us because we have had the necessary insight; to others they must remain meaningless until there has been an equivalent insight. Since this may never occur, the techniques of pointing are relatively unsatisfactory as devices for clarifying hypothetical symbols. Resort to the method of *reduction* is helpful for such clarification, and this is the technique commonly used. If, in the absence of pointing, we wish to make clear to another individual what we mean by molecules, all that we can do is to describe the facts which are explained by this notion; in this way we "reduce" the idea of molecules to the facts, and thus "define" molecules by the facts.

The other problem common to all three forms of science is the problem of truth. For descriptive science truth is a matter of correspondence between a statement and a fact: if there is a fact represented by the statement it is true; if there is no such fact the statement is false. The attempt is made, therefore, to limit the range of statements to facts whose existence can be unequivocally determined. Statements about universal facts, such as that all metals conduct electricity, or that all organisms come from previous organisms cannot be made, since we can never get facts which include *all* possible cases. Statements of limited correlations, or generalizations which merely sum up the cases examined, are the only ones possible. Truth in descriptive science is simply a matter of stubborn fact; things are discovered to be what they are, and there is no question as to why they have the character they do have. A good descriptive science is a large collection of true statements which individually correspond to the individual facts, and collectively correspond to the totality of observed facts.

Truth in formal science is not a matter of correspondence but of deduction. To say that a statement in such a science is true is merely to say that it can be logically inferred from some other statements which are taken to be true. But a strictly formal science has no descriptive reference and therefore cannot be said to be true at all in the usual sense of the word. Hence the notion of truth is replaced by the notion of validity. Neither the postulates nor the theorems are strictly true. The postulates are simply agreed-upon statements, largely arbitrary in character, and therefore not the kind of thing that can be true or false. The theorems, again, are neither true nor false but only correctly inferable from the axioms and postulates or possibly at present undecidable because not yet proven. Truth, therefore, is, like definition, a matter which is internal to the science and determined by the ways in which the symbols are interrelated. A system which is consistent is true in the only sense in which it can be, that is, free from contradiction. Whether or not it talks about the world, whether or not it has any interpretations, are matters which are irrelevant to the science as a purely formal scheme. Universal statements, like those given above, are found in great numbers in such a science. But when we say that they are true we mean only that they are necessary, that is, that they can be correctly inferred from the underived ideas of the scheme. For example, if metals are *defined* as substances which conduct electricity, it can be inferred necessarily that all metals conduct electricity. The law becomes "true" by definition, and it makes no difference whether any *actual* metals do or do not conduct electricity.

Truth in explanatory science is determined by correspondence, by induction, and by confirmation. The statements in the descriptive portion of the science are true by correspondence. To the extent to which idealizations and elaborate measurements have been included the correspondence is less direct, and it may be somewhat difficult to define clearly what is meant by such correspondence. Truth in the case of hypotheses is determined by logical relationships; the statement of the existence of a hypothetical entity is verified by showing that the consequences of this statement are themselves true by correspondence. Thus in explanatory science truth is defined either in terms of correspondence or in terms of implicative relations to other symbols whose truth is determined by correspondence. Take, for example, the generalization, "All cows are ruminants." This is not true by correspondence because we can never find the fact which includes *all* cows. But the statement does imply a series of statements about particular cows, each of which can be shown to be true by correspondence. This permits us to say that the law is true, or, more exactly, highly probable. Truth in descriptive science is a matter of brute fact; in formal science a matter of deductive necessity; and in explanatory science a matter of high probability.

5. Examples of Three Forms of Science

The problem of finding examples of the three forms of science is a difficult one to solve. There are no "pure" sciences of any of the types. Fact-gathering always contains some elements of theory, either as implicit in what are *presumed* to be facts, or as the guiding principle by means of which the facts were originally selected. Pure deductive elaboration never takes place entirely apart from the natural world, since we must think about classes or elements or relations or other abstract structures, and these must exist in the sense already indicated. And explanatory science, being itself a combination of fact-gathering and deductive explanation, necessarily combines both descriptive and formal science, with varying emphasis on the one or the other.

However, approximate or rough examples of the three types can readily be found. In fact, in order to locate descriptive sciences all that we need to do is to refer back to our discussion of the two roles of proper names in science.[12] Unquestionably the best examples of descriptive sciences are the ideographic sciences —geography, astronomy, geology, archaeology, and large portions of the behavioral sciences. These are, of course, ideographic in

[12] See above, pp. 53–56.

varying degrees, and there is no requirement that they be entirely without nomothetic aspects. But in such sciences, as we saw above, individuals and happenings are located in space and time, and they are given proper names. The nomothetic characteristic is introduced since the subject matters of these sciences are not merely named but classified, associated, ordered, and perhaps measured. No laws, in the strict sense, are included; no theories or hypotheses are proposed; and even averages and certain other statistical values have a questionable place. In such sciences we have simply more or less loosely organized collections of described things and happenings. Such a book as the *World Almanac* for any given year would constitute a fairly comprehensive ideographic science. Less sweeping "sciences" would be annual reports of industrial concerns, universities, state and national governments, churches, societies, and so on.

When proper names play their other role in science, i.e., that of temporarily labeling objects for the purpose of generalizing on particular cases and making correlations, they no longer produce purely ideographic sciences. What they create are sciences which are primarily nomothetic, since they contain statements about *kinds* of events and objects, rather than about the individual events and objects themselves, and about *correlations* which are taken to refer beyond the cases actually examined. However, although they are no longer strictly descriptive sciences they fall into a well-recognized category of sciences, commonly characterized in terms of their methods as "background research" and in terms of their results as the "handbooks" and "tables of data" which are so important for all science. Many scientists spend their entire careers gathering and tabulating data of this kind[13] — tables of boiling and freezing temperatures of substances, densities, atomic weights, coefficients of expansion, solubility; tables of temperature, rainfall, wind velocity, hours of sunshine for various locations on the surface of the earth; tables of phases of the moon, positions of the planets, intensity of radioactive fall-out, cosmic ray activity for various places and times; statistical tables used by insurance companies, covering life expectancy, mortality rates for certain common diseases and in certain occupations; and so on through tables giving amount of unemployment, cost of living,

---

[13] I. B. Cohen, *Science, Servant of Man* (Boston: Little, Brown & Co., 1948), p. 53.

prevailing interest rates, rate of population growth, and average income.

At the other end, pure mathematics best exemplifies formal science. Many parts of theoretical physics have been formalized, and even some parts of biology and economics. Celestial mechanics is a formal science while its close relative, astronomy, is primarily descriptive. In fact any collection of data *may* be formalized by means of the idealizing activities, and a *pair* of sciences thus created. Our experience has shown that this formalization is both easier and more useful when we are able to use measuring techniques than when we are not. We profit by this method, therefore, when we are in the area of the abstract and natural sciences (usually called "lower" or "more fundamental") than in the area of the behavioral sciences (commonly called "higher").

The best examples of explanatory sciences are the laboratory or experimental sciences: physics, chemistry, botany, zoology, medicine, geology, and possibly behavioristic psychology. Certainly among these sciences physics and chemistry have the most elaborate theoretical structures, and psychology the least. Whether the behavioral sciences—sociology, economics, political science, psychology—are considered to be explanatory sciences would depend on the degree to which strong emphasis is placed on theory rather than on data obtained through interviews, questionnaires, and other descriptive techniques. To talk about social, economic, or political *theory* over and above social, economic, and political *systems* as found at given times and places, would imply an interest in explanation rather than in mere description.[14]

[14] For an excellent discussion of whether behavioral sciences "explain" see Robert Brown, *Explanation in Social Science* (Chicago: Aldine Publishing Co., 1963).

8

The Scientific Habit of Thought

The attempt has often been made to contrast the personality of art with the impersonality of science by saying that art is *myself* while science is *ourselves*.[1] This presumably does not mean either that art is lacking in universality or that science has no relation to genius. Art does speak to all peoples, and science is the product of such individuals as Ptolemy, Copernicus, Galileo, and Darwin. Yet the artist does have a style which enables us, once we are familiar with it, to identify him by his work, not only as a member of a school but as a specific individual who applies the techniques of the school in a personal and recognizable way; his pattern shows up in his creations. In science this does not seem to be the case. From the fact that Laplace invented the nebular hypothesis we could hardly infer that he must have formulated certain important principles in probability theory; because Newton made discoveries in optics there is no reason to believe that he must have

[1] Claude Bernard, *Introduction to the Study of Experimental Medicine* (New York: Dover Publications, 1957), p. 43.

invented the calculus. Yet art critics can, within a certain margin of error, usually identify the works of Cézanne or of Rembrandt, because the style is indelibly impressed on the product.

This is not surprising. For the creation of the artist is something concrete—a painting, a statue, a symphony, or a poem, while the discovery of the scientist is an abstract theory, not identifiable by lines or colors or meter or rhythm. The artist devises a new art form, often ridiculed and scorned by his contemporaries, and not subject to "corroboration" by any recognized principles. For beauty is in the eye of the beholder. But a scientific theory is public and universal, testable by all who are competently trained to do so. The theory of gravitation, once discovered by Newton, was no longer his own; it belonged to the community of scientists. Genius entered into its creation, but it was accepted because it survived the hard scrutiny of scientific logic. The methodology of science is science in its universality, where rules and principles dominate, and personality—except in rare cases—recedes into the background.

From such considerations as these one would be tempted to draw the conclusion that there is no very significant relation between the personality of any scientist and the scientific studies he is carrying on. Certainly if there were such a connection it would seem to be more remote than that between the artist and his creations. Artists are temperamental, and their idiosyncrasies show up in their work; they often live "peculiar" lives, frequently in active associations only with their own group (all of whom are "peculiar" in the same way). Scientists, on the other hand, at least outside their laboratories, are quite normal in behavior and appearance; they tend, perhaps, toward absentmindedness and a preference for hirsute adornment, but ordinarily one would have great difficulty distinguishing them as a group from doctors, businessmen, lawyers, or politicians.

Yet it is of some interest to note that the scientist in recent years is apparently becoming a type. A study,[2] published recently described the image which college students, and even the scientist's own colleagues, have formed of him. He is highly intelligent, individualistic, uncultured, radical in social outlook, somewhat retiring, generally unhappy in his home life, and (for some in-

[2] D. C. Beardslee and D. D. O'Dowd, "The College Student Image of the Scientist," *Science*, Vol. 133, No. 3457, 997–1001.

explicable reason) married to an unattractive woman. With the accuracy of this image I shall not be concerned, since I cannot conceive that any of these characteristics, with the possible exception of the first, could have any important relation to his competence as a scientist. Presumably the students and professors who provided the data for this image were not asked whether science, being what it is, would be most effectively pursued by a man having these personality characteristics.

This is the question which I should like to investigate. Being suspicious both of opinion polls and of the capacity of students to evaluate scientific skills, I shall draw my material from scientific autobiographies and biographies.[3] Here we have either the scientists themselves speaking, or others who carefully studied their writings and in many cases knew them personally. These sources reveal that there is, indeed, a typical scientist, not in the sense of one who has characteristic social or political views or who has a certain kind of home life, but in the sense of one who exhibits a certain psychological and logical make-up which accounts for his ability to carry on effective research. He must have certain drives which impel him to pursue problems and solve them; he must be painstaking in his observational and experimental methods; he must be imaginative; he must be orderly in his thinking; he must be cautious in generalizing; and so on. These are relevant because there is a pattern of scientific inquiry, and certain qualities of mind are required if this method is to be effectively applied to the objects and processes of nature. P. W. Bridgman said that science is "sciencing," by which he presumably meant that science is in the final analysis the work of persons doing science according to the accepted pattern. There is danger, of course, in talking about *the* scientist or *the* scientific method, which are as artificial as *the* man-in-the-street to whom we so often appeal. Yet I think there is a certain level of abstraction at which we can, with some measure of safety, describe the intellectual qualifications of the

[3] For general readings in this area see: W. I. B. Beveridge, *The Art of Scientific Investigation* (New York: Random House, Inc., 1957), especially Chaps. III, V, VI, and XI; I. B. Cohen, *Science, Servant of Man* (Boston: Little, Brown & Co., 1948); Anne Roe, *Making of a Scientist* (New York: Apollo Editions, Inc., 1953); Anne Roe, "The Psychology of the Scientist," *The New Scientist: Essays on the Methods and Values of Modern Science*, ed. P. C. Obler and H. A. Estrin (Garden City: Anchor Books; Doubleday & Company, Inc., 1962), pp. 82–92; Francis Bello, "The Young Scientists," *ibid.*, pp. 62–81; and J. Bronowski, *Science and Human Values* (New York: Harper and Brothers, 1951).

good scientist insofar as these are relevant to his effective pursuit of science.

I should like to suggest that he possesses certain qualities of mind which can be described in a highly stylized (and to this extent artificial) manner by the terms *fervent impartiality, imaginative objectivity, tolerant rigidity,* and *critical illogicality.* That each of these pairs of terms depicts opposing qualities is apparent. For example, to be impartial requires the scientist to resist the tendency to indulge in wishful thinking; but to be fervent requires him to be intensely devoted to something. How, then, can he be fervently impartial? We may ask the same question with regard to the other attributes. How can he be faithful to the facts while admitting the legitimate role of the imagination? How can he be tolerant of alternatives yet recognize that there is only one truth? How can he accept the logical pattern of scientific thought and yet claim that genius cannot be guided by rules? It appears that the scientist has a very delicately balanced personality, made up of "truces" between these opposing qualities of mind, and when either member of each of these pairs dominates the other the balance is destroyed and something less than good science results. The scientist is either carried away by his enthusiasms or he lacks the drive for effective work; he becomes either a fact-gatherer or a mere speculator; he opens his mind freely to all theories, good or bad, or he closes his mind to all but his own pet ideas; his thoughts are critically centered on the logic of his procedure or he "does what comes naturally" and relies on hunches, intuitions, and unconsciously formed habits. I should like to discuss these paradoxical qualities and see how they are exhibited by the scientist.

1. Fervent Impartiality

The devotion of the scientist to the pursuit of knowledge is too well recognized to require elaboration here. The scientist, perhaps more honestly than anyone else, can say in the words of A. H. Clough,

> It fortifies my soul to know
> That though I perish truth is so.

The theme of scientific martyrdom runs through history, poem, and story. While the actual cases of individuals who gave their lives for the advancement of science are few (Socrates and Bruno are the best-known examples), we tend to forget how many have shortened their lives by long hours in the laboratory, through accidents suffered in the course of research, through illness contracted in the study of disease, through exposure to hardships and privations while engaged in geographical explorations, and so on. A. E. Housman says that the passion for truth is one of the weakest of human passions. This seems to me to be contradicted by the history of science. The passion for truth may be the *rarest* of human passions, since the number of scientists and scholars is small compared to the total population, but it certainly is not the *weakest*. T. H. Huxley, in a letter to Charles Kingsley, described this devotion to truth, which is so characteristic of the scientist:

> Science seems to me to teach in the highest and strongest manner the great truth which is embodied in the Christian conception of the entire surrender to the will of God. Sit down before fact as a little child, be prepared to give up every preconceived notion, follow humbly wherever and to whatever abyss Nature leads you, or you shall learn nothing.[4]

Whether this overpowering concern for truth is motivated simply by the honest and sincere desire to know, or by the more practical urge to improve man's lot on earth really makes no significant difference. Many feel that only theoretical truth is pure, that knowledge which is applied to concrete problems becomes thereby polluted and defiled. Others insist, with Goethe, that while the world is green all theory is gray, or, with Bacon, that knowledge should be pursued only because it provides us with power over nature. The history of science abounds in examples of these two extreme types, and there seems to be no reason for identifying "genuine" science with either. In both cases there is that wholehearted devotion to knowledge without which science could not exist; but how this is produced in the mind of the scientist is of no critical importance.

This is, however, only one side of the scientific personality. While a strong dedication to truth is essential to science, an uncontrolled enthusiasm can bring about its destruction. An ardent love can be blind in matters of the intellect as well as in human relationships. Extreme devotion to any cause may produce an

[4] T. H. Huxley in a letter to Charles Kingsley. *Science in a Changing World*, ed. Mary Adams (New York: Century Co., 1933), p. 230. Reprinted by Permission.

intellectually warped vision. Bacon said that the scientist must never show eyes lustrous with human passion. Sinclair Lewis, in *Arrowsmith*, states what he calls "the prayer of the scientist":

> God give me unclouded eyes and freedom from haste. God give me quiet and relentless anger against all pretence and all pretentious work and all work left slack and unfinished. God give me a restlessness whereby I may neither sleep nor accept praise till my observed results equal my calculated results or in pious glee I discover and assault my error. God give me the strength not to trust God.[5]

In other words, may my fervent pursuit of truth be tempered by the realization that there are no short cuts to knowledge, that "pet" theories are dangerously enticing, and that the worship of truth may unconsciously become idolatrous! Huxley said that science commits suicide when it adopts a creed, and he also remarked that there is nothing like a sordid fact to slay a beautiful theory. "Truth," said Renan, "is a great coquette; she will not be sought with too much passion, but often is most amenable to indifference. She escapes when apparently caught, but gives herself up if patiently waited for; revealing herself after farewells have been said, but inexorable when loved with too much fervor."[6]

2. Imaginative Objectivity

Here the problem of the personality of the scientist is that of reconciling two strong, but often sharply opposed, urges—one, to remain faithful to the facts and to avoid fruitless speculation, and the other, to explain and account for the facts which have been disclosed through observation and experimentation. The proper understanding of science clearly involves recognition of the respective roles of both data and hypotheses, and the problem of uniting them in the total picture is rendered not only insoluble but utterly nonsensical if it is stated in terms of the "relative importance of facts and ideas in science"—a formulation which reminds us of some of the topics which we used to propose for high school debates—"Resolved that men are more important than women for the propagation of the human race." There are,

[5] Sinclair Lewis, *Arrowsmith* (New York: Harcourt, Brace & World, Inc., 1925), pp. 280–281. Reprinted by permission of the publisher.

[6] From J. E. Renan's speech nominating Pasteur for membership in the Académie Française. René Radot, *Life of Pasteur* (London: Constable & Company, Ltd., 1902), II, 164.

of course, no ideas without facts and no facts without ideas. Darwin said that no one can be a good observer unless he is an active theorizer,[7] but he also said that any fool can generalize and speculate.[8] Bacon spoke of the need for "holding the imagination in leash"—a particularly apt phrase since the scientist commonly feels the "pull" toward venturing an explanation before adequate foundation has been laid in the facts. Buffon said that genius is essentially patience; Tyndall, as we have seen,[9] stressed the need for "preparing" the imagination by carefully examining the facts; and Santayana coined the phrase, "chastened imagination."

But since the mental qualifications of a good observer are in some respects quite different from those of a good theorizer there is a strong tendency to split science into two corresponding domains, each of which has its devotees who promptly invent derogatory terms to characterize the members of the opposing school. On the one hand are the *accumulators*, the fact-gatherers, the pebble-pickers, the gadgeteers, and the bird-watchers; on the other hand are the *guessers*, the dreamers, the fanciers, the speculators, and the woolgatherers.

The accumulators are those who love facts and hate speculation and theorizing; the guessers are those who are bored by mere facts but tremendously excited by sweeping generalizations and integrating ideas. The former are represented by Wordsworth's poem, *Peter Bell*, for whom

> A primrose by the river's brim
> A yellow primrose was to him,
> And it was nothing more.

The latter are represented by Tennyson, who felt that the flower in the crannied wall, if we could understand it "root and all, and all in all," would tell us what God and man is. Certainly there are good scientists of these extreme types, and it is ridiculous to insist that there should be no scientists who speculate, or no scientists who merely gather data. Perhaps May Kendall, when she said of Thomas Huxley[10] that

---

[7] *Life and Letters of Charles Darwin*, ed. Francis Darwin (New York: Basic Books, Inc., 1959), I, 126.

[8] Frank Cramer, *The Method of Darwin* (Chicago: A. C. McClurg & Company, 1896), p. 39.

[9] See above, p. 97.

[10] *Life and Letters of Huxley*, ed. Leonard Huxley (London: Macmillan & Co., Ltd., 1900), I, 112. Reprinted by permission.

Primroses by the river's brim,
Dicotyledons were to him,
And they were nothing more,

was trying to put him mid-way between the accumulators and the guessers. Once the artificiality of the separation of facts from ideas has been recognized, the dichotomy ceases to have meaning and the problem becomes one of indicating the respective roles of data and hypotheses in the total scientific procedure. Darwin, for example, stated that he had an unusual capacity for observing facts. Yet he admitted that obvious data often completely escaped his notice. On one occasion[11] when he was looking among rocks for fossils he was so intent on what he was searching that he failed entirely to notice the evidence of glaciers—the plainly scored rocks, the perched boulders, the moraines; yet these phenomena were so conspicuous, as he said later, that a house burnt down by fire did not tell its story more plainly than did these facts. On the other hand, Faraday was strongly imaginative. Tyndall says of him:

> When an experimental result was obtained by Faraday it was instantly enlarged by his imagination. I am acquainted with no mind whose power and suddenness of expansion at the touch of new physical truth could be ranked with his. Sometimes I have compared the action of his experiments on his mind to that of highly combustible matter thrown into a furnace; every fresh entry of fact was accompanied by the immediate development of light and heat. The light, which was intellectual, enabled him to see far beyond the boundaries of the fact itself, and the heat, which was emotional, urged him to the conquest of the newly revealed domain. But although the force of his imagination was enormous, he bridled it like a mighty rider, and never permitted his intellect to be overthrown.[12]

To be faithful to the facts, yet to see behind the facts—this is the essence of imaginative objectivity.

3. Tolerant Rigidity

Openness of mind is an important attribute of the scientific thinker. New theories—indeed, revolutionary theories—are continually emerging from the background of research. While there is con-

[11] Cramer, p. 129.

[12] John Tyndall, *Faraday as a Discoverer* (New York: Appleton, 1898), p. 108.

siderable exaggeration in the reply of the physicist who when asked what the latest theory of matter was replied that he could not answer the question since he had not yet seen the morning paper, there is a measure of truth in what he said. What should be the attitude of the scientist toward each new theory which presents itself? Since theories arise through imaginative insight and this method by itself provides no criteria for validation, the scientist can neither accept nor reject a theory until satisfactory confirmation or disconfirmation has been provided. But here is precisely where the difficulty arises. For how conclusive must the confirmation be before a theory is admitted into the body of science and how conclusive must its disconfirmation be before it is discarded? Ernest Nagel[13] has discussed the problem of degrees of confirmation as applied to hypotheses. He points out that no quantitative values can be assigned to them and that matters of mere "convenience" often enter into our decisions. Thus there are no logical principles on the basis of which we know when to accept one theory and when to reject another. The geologist, T. C. Chamberlin, proposed a method which he called that of "multiple working hypotheses."[14] While carrying on an investigation, he suggested, we should have in mind not a single hypothesis which we are trying to confirm but a wide range of alternative hypotheses, so that evidence which fails to bear on one will bear on another and thus will not escape notice entirely. This seems to characterize that openness of mind which the scientist brings to his study.

But a mind can be admittedly too wide open. The scientific discipline is severe and its criteria of validity are high. Both the facts and the methods of science have been won through the sweat and blood of centuries of disciplined effort, and the scientist has a genuine respect for them. Perhaps he is justifiably impatient, therefore, with circle-squarers and makers of perpetual motion machines. Teachers of high school and college mathematics are plagued, I am told, with students who claim to be able to prove that 1 equals 0. Certainly the scientist is not fairly to be charged with having a closed mind when he is unwilling to

[13] Ernest Nagel, "Principles of the Theory of Probability," *International Encyclopedia of Unified Science*, I, No. 6, 66–75. Also see above p. 128.

[14] *National Academy of Sciences: Biographical Memoirs* (Washington, 1934), XV, 388; see also Cramer, pp. 40–43.

listen to those who insist that they have made great discoveries but who have never learned the primer of science. A common complaint of scientific quacks is that scientific journals will not publish their manuscripts and established scientists will not listen to their arguments. But being openminded does not mean being tolerant of all solutions which are offered for scientific problems, or of solutions proposed for problems which are known to be insoluble. Some things which pose as science are just plain wrong and the scientist can hardly be criticized for being intolerant of this sort of thing. One who does not know what has been done in science cannot expect to make contributions to science. Faraday complained that he was continually bothered by novices who had "discovered old facts."

4. Critical Illogicality

The problem of this final attribute of the scientific personality is simple, yet it is a troublesome one. No one, I should suppose, would ever question the importance to a scientist of a knowledge of logic. Science, in fact, *is* logic applied to a certain subject matter. But this is not precisely the issue. The problem is, rather, to determine the role of conscious logical knowledge in the actual procedure of science. To what extent should the scientist think about his logic while he is carrying on his investigations?

On the one hand, some knowledge of the rules of thought seems essential. The scientist must know how to draw valid inferences, how to make accurate observations, how to measure and experiment, and he must certainly know something of the subtle way in which prejudice, emotion, superstition, and unconscious philosophies of life creep into scientific investigations and warp scientific conclusions. The scientist who knows something of logic is, therefore, other things being equal, more likely to contribute to good science than his colleague who does not. Without the knowledge he is more prone to error, more easily confused by complex problems, and less able to expound his results convincingly to others.

But there is much to be said on the other side. Many scientists scorn courses in logic and ridicule attempts to direct the operation of the scientific method by "laws of thought." Science, they would claim, just goes on; it is spontaneous and self-regulative; it proceeds by hunches and is based on well-formed habits acquired during the training period. Someone recently defined an expert as one who does not know what he is doing. In a certain sense this is what a good scientist is: he is like a skilled automobile driver who makes his decisions automatically and does not have to deliberate whenever an unexpected situation arises. To be sure, a scientist must have acquired this skill at some point in his education; but he learned the scientific method by practicing it under a skilled scientist, not by taking a course in logic or by reading a book which analyzes the scientific method. However, once the skill has been acquired, according to these advocates, he should drop it from reflective consideration and allow it to play its true role as the unconscious motivator and regulator of the scientific knowing process. Indeed, the study of logic may be not only useless, but positively harmful. The scientist who thinks continually about what goes on in his head is likely to fail to see what takes place before his eyes. Or the scientist who is continually *thinking about* his method may find that he can no longer *use* the method. In this regard the scientist is much like the centipede who was asked how he could manipulate so many legs with such perfect coordination. He replied that it was all very simple; in fact, just like this—and then when he attempted to *analyze* his complex behavior he found that he could no longer *perform* it, and he was never able to walk from that time on.

There seems to be both truth and error in this claim. The truth is that too much concern for the thought processes while they are going on does interrupt their smooth operation. The error lies in the assumption that the skill of the scientist necessarily manifests itself in this way. Logic contributes to thinking, not while it goes on but after it has been completed; logic acts as the critical evaluator which tells us what kinds of error there are and how to guard against them. From this point of view the scientist can hardly know too much logic. In his use of methodological considerations, therefore, the scientist may well think rationally without necessarily destroying spontaneity. Probably no scientist can proceed effectively without a rudimentary knowledge of the distinction between fact and theory, between valid and invalid inference, between experiments which are properly controlled and those which are not. To this extent logic and the reasoning processes are needed in science. But logic does not *make* science; it gives direction to science but cannot be a substitute for it. The good scientist is one who recognizes the merits and limitations of logic; he knows the logic of his procedure but he also knows that there are not any "rules," as Bacon evidently thought there were, by which the mind can extract ideas from nature in much the same way that one can extract juice from an orange. Genius does not obey laws; it makes them.

9

Scientific Truth as a Value

The subject matter of our study now changes in character. We have been concerned with the attempt to make clear what is involved in the scientific method, which we characterized in its complete form as the "method of confirmed hypothesis." We recognized the limiting cases of descriptive sciences, which tend to avoid hypotheses, and of formal sciences, which refer only to the abstract aspects of the world. We suggested that science is a special type of investigation in a more general field which can be called "inquiry," but we were not able to go into further detail in the exposition of the meaning of this term. Our task will now be fourfold: first, to determine whether science is concerned with value; second, to explain what is meant by preferential behavior; third, to attempt to make a new distinction between science, on the one hand, and the humanities, arts, and philosophy, on the other; and, fourth, to show that science is a value pursuit essentially like art, religion, political behavior, work, and the other characteristically human responses to the world.

1. Science As Non-Valuational

The topic can be best introduced by indicating a point of view[1] which, although containing a measure of truth and actually favored by tradition, may lead to a basic misunderstanding of the nature of science. Science and value, according to this conception, are incompatible. Any interest which is scientific can have nothing to do with values, and any concern with values must lie outside the realm of science. Consider, for example, the scientist as a personality. He is, as we saw in an earlier chapter, the ideal exemplification of objectivity and unemotionality. Likes and dislikes, hopes and fears, preferences and aversions, all of which create value attitudes, do not enter into his pursuit of science. He spurns wishful thinking, and is justifiably suspicious of any of his colleagues who appear to "juggle the facts" in order to bring them into agreement with a favorite theory. He may willingly grant that the pursuit of beauty, goodness, piety, and happiness are legitimate activities of the human spirit, but he knows that such pursuits are not science, and are therefore of no concern to him *insofar as he is living his role as a scientist.* Values are relative to the individual, personal, and private. Truth, on the other hand, is objective, impersonal, and public. Good and bad are such only because "we think so"; there would be neither good nor evil in a universe in which there were no human beings. Beauty is believed to be dependent on the beholder. Even God is demanded to satisfy our feelings and not to meet any intellectual needs. Science will have none of this. It is not *moral*; it is not *immoral*; it is *amoral*—entirely outside the field of value judgments. Its seed is fact; its nourishment is careful observation and cautious reasoning; and its flower is pure truth—whether we like it or not.

The force of this argument is considerably weakened when we

---

[1] There are many good references on this topic. Some of the best are the following: J. Bronowski, *The Common Sense of Science* (Cambridge: Harvard University Press, 1958), Chap. VIII; George Boas, "The Humanities and the Sciences," *The New Scientist: Essays on the Methods and Values of Modern Science,* ed. P. C. Obler and H. A. Estrin (Garden City: Anchor Books; Doubleday & Company, Inc., 1962), pp. 178–198; E. A. Burtt, "The Value Presuppositions of Science," *ibid.,* pp. 258–279; G. A. Lundberg, *Can Science Save Us?* (New York: Longmans, Green & Company, 1947), Chap. V; Moody E. Prior, *Science and the Humanities* (Evanston: Northwestern University Press, 1962), Chap. I; H. G. Cassidy, *The Sciences and the Arts* (New York: Harper and Brothers, 1962); Bernard Barber, *Science and the Social Order* (Chicago: The Free Press of Glencoe, 1952), Chap. XI; and *Science and the Creative Spirit,* ed. Harcourt Brown (Toronto: University of Toronto Press, 1958).

recognize two facts. The first is that even if the disregard for values applies to science it does so only to *pure* science, not to *applied* science. Practical science is science which is used in the world for the purposes of making us more comfortable, healthier, richer, more ethical, more appreciative of beauty, and even more religious. Oppenheimer, director of the Los Alamos atomic bomb project, is reported to have said after Hiroshima, "Science has at last known sin." Here the relevance of science to values is obvious, for comfort, health, money, virtue, beauty, and piety are clearly things which we desire, and which are therefore expressive of our preferential reactions to the world in which we live. Applied science, in this sense, can be considered to be one of the useful arts or one of the humanities; it concerns itself with the attempts which man has made in his long history, and is making today, to increase human happiness and to weaken or destroy the forces causing discomfort, disease, poverty, evil, ugliness, and sin.

The second fact which contradicts the claim of science to be unconcerned with values is that the behavioral studies are commonly admitted among the sciences. This is not, of course, universally the case. Whether sociology, political science, economics, and psychology — to take the most obvious cases — are to be characterized as sciences depends on how "science" is defined in the first place. According to the definition already given,[2] the behavioral sciences fall into this category. They tend to be descriptive in character rather than explanatory, though this varies among the individual sciences; they are not formal sciences, except in certain limited areas of their subject matter; they tend to be strongly ideographic rather than nomothetic.

According to the two very rough definitions of "science" — on the basis of whether or not the methods employed by the investigator manifest the scientific spirit, and according to whether they are dominantly quantitative and experimental — the cultural sciences do not fare too well. Wherever man is studying man, objectivity is hard to achieve; to be objective one must be, so to speak, outside his subject matter and neutral toward it. But when that subject matter is human hopes, pleasures, fears, disappointments, and enthusiasms, either as described by the subject or as exhibited in less direct behavioral ways, the investigator can hardly remain wholly objective. To be human is to be sympathetic toward other

[2] See above, p. 20.

human beings; to be a scholar of human behavior is to report it without either sympathy or antipathy. Thus the behavioral scientist is either a human being or a scholar, but he cannot be both. To state the problem this way is, of course, to reduce it to an absurdity. No scholar is wholly neutral toward his subject matter though he finds it much easier to be so if he is a physicist than if he is a psychologist or a sociologist. All that we have to do, therefore, in order to admit behavioral studies into the field of legitimate scientific enterprises is to acknowledge frankly the special difficulties to which the investigator in this area is subject, and to allow him to handle them for himself as ably as he can.

The same point of view applies to the inability of the behavioral scientist to use quantitative and laboratory methods in the behavioral sciences. Neither the behavior of individuals themselves, nor the factors which motivate them and result from this behavior can readily be quantified; as we saw in our discussion of measurement, occasionally we can draw up in the behavioral studies crude preferential scales, to which numbers may be more or less arbitrarily attached. But this is a very loose and rough form of quantification. As for experimentation, while this does occur in the behavioral sciences, as we have seen,[3] strict control of conditions is impossible because of the complexity of the variables entering into the problem. A much more important difficulty is that experimentation on human beings requires the consent and cooperation of those who are to be the subjects of the experiments. But most individuals, even though they fully appreciate the value of scientific knowledge, are unwilling to subject themselves to living under the unusual and sometimes disagreeable conditions set by the experimenter. I think it was William James who at one time said that the experimental social scientist (and this applies also to the biological scientist) should prepare himself for the time when his experimental subject will suddenly raise his head—if he *has* a head—and say, "Look! It is *me* that you are experimenting on!" This possibility will prevent the cultural sciences from ever using experimentation to any great degree. But it will not prevent them, nor will any other of the limitations mentioned, from being generously included among the behavorial sciences, as has been the case throughout our study up to this point.

[3] See above, p. 37.

Now if we correct this argument for the non-value character of science by admitting that *applied* science certainly has much to do with values, and that *behavioral* science, in its concern with human living, is shot through and through with choices and preferences, the question as to whether science is non-valuational cannot be answered affirmatively without further examination. To solve the problem requires us to start with something which seems quite remote from science, namely, living itself.

2. Human Living As Preferential

When we ask what this business of living involves, the answer at which we arrive (after due deliberation) seems quite clear. Living —human living—is characterized essentially by the making of choices and acting as a result of these decisions more or less freely made. Man, in contrast to the lower animals, and in contrast to inanimate nature, is a *purposeful* or *goal-seeking* being. By this is meant simply that he is a creature with conscious likes and dislikes, attractions and aversions, who is so constituted as to be happiest when he is seeking and finding those things from which he derives pleasure and when he is avoiding those things which cause him pain or displeasure. His life, in other words, is built around the attempt to render certain positive values as lasting and intense, and certain negative values as transitory and weak, as his power and the constitution of the universe permit. Just what this activity involves will be clear, I hope, in what follows. Sometimes, as in the case of stable and well-organized individuals, it is characterized by persistence and oneness of purpose; in other cases, it is manifested by interests which are often chaotic and confused. Even in the best-oriented individuals the activity inevitably involves conflicts among the preferences, and the result is the sacrifice of certain goals for the attainment of others.

If, granting this fact, we attempt to characterize man in terms of what seems to be his most significant activity we may call him a "preferential animal." He makes choices which are frequently more or less automatic, sometimes conscious but based upon very little deliberation, and at other times the outcome of careful thought and premeditation. The making and result of such choices is behavior, and this is substantially what is studied in the behavioral sciences. Of course man, as a part of the physical world and as a member of a society, is often compelled to behave in ways contrary to his decisions and in ways which have been preceded by no relevant decisions. The behavioral sciences, therefore, must include within their subject matter both preferential behavior and that behavior which is the result of illness and accidents, the forces of nature, and physical acts of other individuals, such as being thrown into jail or being kidnapped, and so on. Here man acts not preferentially but as a result of forces not under his control; but these actions are comparatively rare in the life of the average man and we shall not violate the common usage of the word "behavioral" if we include them in the subject matter of these sciences.

One of the most transparent facts about preferential behavior is its manifestation in two roughly differentiated forms, determined by whether the choices involved are trivial, superficial, and transitory, or important, more profound, and more lasting. In the former group are found one's preference of ice cream to chocolate cake, of cigars to cigarettes, of golf to tennis, of urban life to rural life, of red neckties to blue ones, of movies to television, and so on; in the latter group are found not only one's preference of food to starvation, of health to illness, of sociability to the life of the hermit, of good government to bad, of occasional recreation to a life of constant work, of knowledge to ignorance, but also one's preference of Democracy to Communism, of a self-centered life to a life devoted to serving mankind, of creative art to business, and so on. The main characteristic of the superficial preferences is that they are so diverse as to forbid subclassifications; they seem to be related in no significant way to the basic personality of the one who makes the choices; they commonly change from day to day; and they produce no profound effect on the individual in case he is unable to satisfy them. The significant property of the more basic choices is that they generally are capable of rough classification into *kinds* of preferences; that they are significantly related to the personality of the individual; that they remain rather constant for the person after he has reached maturity; and that they tend to produce disappointment, frustration, and unhappiness if they cannot be achieved at least to a degree.

It is with these more basic preferences that the behavioral scientist is primarily concerned, for when he examines human living he finds that they are surprisingly few in number, capable of more or less sharp differentiation from one another, and provide a basis for breaking up behavior into special fields. Each of the basic preferences provides a center around which activities and experiences group themselves and attain a certain unity, and the behavioral scientist thus has at hand a subdivision of human behavior into special areas, determined by the goals involved. Let us consider some examples.

One of the recognized behavioral sciences is political science. It is concerned fundamentally with all of the choices, acts, and experiences of individuals in their attempt to set up and maintain good government. It explores such concepts as the individual, the State, citizenship, sovereignty, law, types of government, representation, power, finance, and international relations. These are all areas within which the individual and the group make choices based on preferences, and then act on their choices by modifying the personnel and structure of society in the hope of attaining a desirable form of government. In much the same way we can define the subject matter of sociology as all of those choices and actions which are designed to promote the happiness of man's life among men, but which often — for reasons which the sociologist hopes to determine — fail to achieve the desired end. Economics is, similarly, a study of the acts and experiences through which individuals and the group pass in their desire to attain the maximum amount of goods and services through the instruments of barter and exchange; the activities we call "work," the goal we call "wealth." But man also wants to play, to be healthy, to create and appreciate beauty, to live an upright life, to be pious, and to be wise. These are behavioral areas which are well defined. Morality, for example, is the sum total of the experiences through which he passes, in his attempt to achieve an ideal or standard of "right living." Art, similarly, is a name for those phases of human behavior and enjoyment which center about the actualization, either in the form of creation or in the form of appreciation, of the value of beauty. Religion, though much more complex than morality and art, and exhibiting itself in a wider range of diverse forms, is the sum total of activities and experiences designed to create an attitude of piety and spirituality in the life of man. Included among such activities are prayer, worship, meditation, sacrifice, helping one's neighbor in distress, proselytizing, and many other varied rites, ceremonies, and behavioral manifestations. Piety could be loosely defined as a way of looking at the world and man which provides comfort in times of privation and sorrow, affords him something toward which he may practice humility, and inspires him to continue in his fight against the forces of evil. These are behavioral areas each of which includes its own clusters of preferences, acts of decision, and intellectual and emotional experiences which are bound together by their common end, but which differ from those defined by alternative values. And each such group of preferences determines a distinct field of study whose task it is to classify, associate, order, and measure (if possible) the behavioral material, and perhaps even to develop hypotheses and theories in terms of which this material can be explained.

The problem of merely listing, not to mention classifying, the behavioral studies is a difficult one and cannot be examined here. Great controversy exists as to what is and what is not a behavioral science, and disagreements are common with regard to how the behavioral sciences, once they have been collected, are to be grouped and subdivided. Charles A. Beard,[4] for example, includes the following: history, political science, economics, cultural sociology, and geography. Even this simple list provokes objections. Some historians, for example, like to be considered as scientists; others as humanists. Human geography is behavioral, physical geography is not. Psychology, which would seem to be the behavioral science *par excellence*, is often grouped neither with the social sciences, nor with the humanities but with the biological sciences. Education may or may not be considered a behavioral science. Is jurisprudence a science or is it not? Is medicine behavioral or biological? Recent years have witnessed the birth and rapid growth of such studies as information theory,[5] cybernetics,[6] linguistics and sign-behavior,[7] decision-making theory,[8] and game

[4] Charles A. Beard, *The Nature of the Social Sciences* (New York: Charles Scribner's Sons, 1934).

[5] Amiel Feinstein, *Foundations of Information Theory* (New York: McGraw-Hill Book Company, 1958).

[6] Norbert Wiener, *The Human Use of Human Beings: Cybernetics and Society* (Boston: Houghton Mifflin Company, 1950); and W. Ross Ashby, *An Introduction to Cybernetics* (New York: John Wiley & Sons, Inc., 1956).

[7] See footnote, p. 45.

[8] R. T. Livingston and D. B. Hertz, *Decision Theory*, Papers 52A–106 (New York: American Society of Mechanical Engineers, 1952); and Paul Wasserman, *Decision Making: An Annotated Bibliography* (Ithaca: Cornell University Press, 1958).

theory;[9] whether these will be listed among the behavioral sciences of tomorrow will depend on how quickly they define their respective fields and set up methods of study appropriate to their subject matters. Most basic of the problems concerning the behavioral studies is that which revolves around the attempt to differentiate between the *scientific* study, and the *philosophical* study of behavior, and to this we now turn.

3. Humanities, Philosophy, and Meta-science

For convenience in trying to distinguish between the more or less descriptive study of preferential behavior and the theoretical, self-critical, and humanistic approach, I should like to propose that we consider the kinds of preferential behavior as exhibiting themselves in the life of the individual and the group on two roughly distinguishable stages of critical awareness.

On the primitive stage they are almost wholly unconscious and unreflective. This is the level not merely of childhood, but also of many adults who approach human living uncritically, accept what falls to their lot, and make no attempt to better either themselves or the world by increasing the amount or availability of health, recreation, labor, sociability, better government, virtue, piety, beauty, or information. At this stage man's health behavior is limited to such simple activities as proper eating, indulging in recreation, and protecting himself against accident and disease. His recreational activities are for fun rather than for utilitarian ends. His working life, commonly looked upon as one of the disagreeable necessities of being human, is made as pleasant as his capacities and the economic system of which he is a part permit. Sociability is simply making friends, enjoying his family, and possibly making minimal contributions to community activities. Governmental service is paying taxes (which he *must* do), keeping himself informed on public affairs and voting (which he usually does not do), and obeying the law. Morality is his automatic yielding to the dictates of conscience or training. Art is an immediate and unanalytic enjoyment. Religion is a spontaneous and unquestioning devotion to something regarded as having

---

[9] J. von Neumann and Oskar Morgenstern, *The Theory of Games and Economic Behavior* (Princeton: Princeton University Press, 1944).

enduring value. And inquiry is likely to be the undisciplined and uncritical acquisition of the minimum of behavioral responses enabling him to "get about" in the world.

On the secondary stage, which is distinguishable only in degree from the primary stage, the preferential behavior becomes consciously recognized, becomes describable, and begins to exhibit internal conflicts between preferences. Decisions are less spontaneous and become more deliberate. Intellect plays an increasingly important role. The effort is made not merely to choose, but to make *wise* choices — choices which will not later be regretted when their full consequences have been revealed. Here art is no longer a mere communication of feeling but demands knowledge of idea, expression, and form; religion ceases to be childlike and asks for creeds and theologies; morality becomes conscious of standards of behavior, and calls for the resolution of moral conflicts — conflicts between duty and pleasure, between immediate and future satisfactions, and between social and individual values; science begins to formulate its methods and to justify its conclusions by the rules of argument; political behavior turns its attention increasingly to ideologies; and so on.

Although we have spoken of the study of this preferential behavior and the values involved therein as "behavioral science," this terminology is not by any means universally accepted. Many scholars insist that the study of certain kinds of preferential behavior should be called "philosophical studies," "normative studies," or the "humanities."[10] This is particularly true of the study of moral behavior (ethics), of art appreciation and creation (aesthetics), of religious behavior (philosophy of religion), and of scientific behavior (philosophy of science). The fact that such studies are commonly taught in colleges and universities in departments of philosophy or divisions of the humanities rather than departments of social (behavioral) sciences is an indication of the acceptance of this usage. Argument for this terminology could be justified by insisting that moral, art, religious, and intellectual behavior, and the goals which they pursue, are somehow more "basic" in the life of man than social, economic, political, health, and play behavior and the goals which they pursue. Reasons for such distinctions could be found in the higher degree of self-

[10] G. C. Homans, "The Humanities and the Social Sciences," *American Council of Learned Societies News Letter*, XII, March 1961, No. 3.

criticism and examination of presuppositions which seems to be demanded by the former when they are compared to the latter. But while the curricular pattern of courses would justify this distinction, the courses as actually taught would not. For example, many courses in anthropology, cultural sociology, and "descriptive morals," which would presumably be behavioral sciences,[11] contain strong emphasis on ethics and the humanistic implications of moral behavior; but many do not. Again, many courses in the history of art introduce art criticism and aesthetics; and many courses in aesthetics describe art objects and art behavior, and relate artistic creations to the science of art materials—pigmentation, the color spectrum, the strength of building materials, acoustics, wave lengths of musical tones, and so on.[12] Still again, many courses in the history of religion and in comparative religion refer to the philosophical and humanistic issues involved; while others are merely descriptive of sociological behavior.[13] Where the courses are taught makes no essential difference; how they are taught, and by whom, are the essential desiderata, since all that matters is the clear awareness on the part of the instructor of the manner in which he is approaching his subject matter, and of the problems which he considers suitable for this sort of study.

I should like to propose a terminology which may serve to clear up some of the ambiguity and confusion which has arisen as a result of the attempt to distinguish between the scientific studies of human behavior and the more humanistic and philosophical studies. As an approach to the clarification I should like to mention a term which arose in connection with certain studies of linguistics and logic. Let us take a simple example. We have found that when we talk about a language we must distinguish between the language we are *studying*, and the language which we are *employing* in this study. If we are teaching French to English-speaking students we use English to talk about French. English then becomes a language about a language, or a second-level language. For instance I may state (in English) that in French the

[11] For example, W. E. H. Lecky, *History of European Morals from Augustus to Charlemagne* (New York: D. Appleton & Co., 1877).

[12] For example, J. Redfield, *Music—A Science or an Art* (New York: Tudor Publishing Co., 1926).

[13] At the December, 1963, Annual Meeting of the American Association for the Advancement of Science, the Society for the Scientific Study of Religion was admitted as an affiliate member.

gender of a noun has no relation to the sex of that which is referred to by the noun. Here French would be a first-level language and English would be a second-level language. Or, suppose I were to say, " 'Paris is the capital of France' is a true statement." Here I have made a statement about a statement, and this would be a second-level statement about the statement given in single quotes. By common consent languages about languages are now generally called "meta-languages" [14] – with or without the hyphen. Languages about the world are called "object-languages" and languages about such languages "meta-languages." Of course there may also be still higher level languages, since we may talk about meta-languages. The conception of meta-languages has also done much to avoid some of the confusions of logic. For example, we can say that "cat" is a word and is not a cat; but we cannot say that "word" is a word and is not a word, for this would be self-contradictory without the notion of a meta-language.

But if there are languages about languages why should there not be sciences about sciences, or, more generally, studies about studies? In each case the second-level investigation would be a meta-science or a meta-study. The terms "meta-logic" and "meta-mathematics"[15] have come into common usage. The term "meta-science" is not so generally employed, though at least one recent writer has incorporated it into the title of a book.[16]

My proposal is that we make the same distinction between science and meta-science that we make between language and meta-language. Then the behavioral studies will constitute what are generally called the "behavioral sciences." As we have seen, these may be restricted to the traditional sciences of sociology, economics, political science, history, geography, and psychology. Or they may be extended to include certain studies of doubtful scientific status, such as jurisprudence, education, and the science of recreation. Or, again, they may be further broadened to include information theory, cybernetics, linguistics, and the other embryonic sciences which are appearing on the scene. Or, finally, they

[14] See, for example, I. Copi, *Symbolic Logic* (New York: The Macmillan Company, 1954), pp. 186–189; and R. Carnap, *The Logical Syntax of Language* (London: Kegan, Paul, Trench, Trubner and Co., Ltd., 1937), p. 9.

[15] Ernest Nagel and J. R. Newman, *Gödel's Proof* (New York: New York University Press, 1960), pp. 28–32.

[16] Mario Bunge, *Metascientific Queries* (Springfield, Ill.: Charles C. Thomas Publisher, 1951).

may be rendered sufficiently inclusive to cover descriptive ethics, history of religion and comparative religions, history of art and art objects, and even the descriptive science of science itself.

But whatever may be included, the meta-behavioral studies will constitute a distinct, though closely related, group. Their main characteristic will be that they are meta-behavioral studies *of* behavioral studies. To characterize them in this way means that the meta-behavioral studies must not be *confused* with the behaviors which are studied, but it does not necessarily mean that the former are "on a higher level" or "more basic than" those which are being studied, nor that the two are generically different. For example, there are many studies such as the psychology of science, the sociology of science, the economics of science, and even the geography of science, which are meta-sciences in the sense that they are meta-behavioral studies of scientific behavior. Presumably when we study, say, the psychology of science, we more or less isolate one aspect of scientific behavior—how the scientist "thinks"—and study this by the methods which are commonly employed by the psychologist whenever he studies any "thinking." The same interpretation could be given to the economics of science and the others mentioned, possibly including the history of science as well if one is not averse to considering history a behavioral science. Thus there seems to be no good reason for setting these sciences apart and putting them into a special category which will distinguish them sharply from the behavioral sciences in general. Consequently I shall not henceforth call such scientific behavioral studies of science "meta-sciences."

On the other hand, the term "meta-science" may prove to be a valuable linguistic tool to characterize certain second-level studies whose methods may be not only quite different from the methods of the sciences which are being studied but also unlike the methods employed by any of the sciences. In such cases meta-sciences might very well be generically different from the sciences which they study—different in method, in attitude toward subject matter, in search for presuppositions and unconscious assumptions, in widening the area of the field being investigated, and so on. Let us examine this problem by means of examples.

Behavioral scientists frequently attempt to distinguish between social science and social philosophy, the science of economics and the philosophy of economics, political science and political philosophy. Presumably by the term "philosophy" in this connection the scientists mean a kind of study which is in some very fundamental way different from the corresponding sciences. In fact they would generally be understood as trying to indicate that a subject matter may be studied scientifically *or* philosophically. The distinction would be very unfortunately described if they were to say that the former was "superficial" while the latter was "profound," or that the former was "merely descriptive" while the latter was an attempt to arrive at "true understanding." Such terms would be highly emotive in character and would readily lead to the conclusion that a science is an inferior and unimportant kind of study while a philosophy is always a superior and much more significant mode of investigation. This is certainly not the case—at least in the absence of any reference to the purpose for which the investigator is carrying on his studies. If he wants scientific knowledge philosophical knowledge will not take its place, nor conversely. In certain contexts science will be more important than philosophy, in others philosophy will be more important than science, and in still others both may be needed.

I think that the issue can be clarified by a judicious definition of the term "meta-science." I should like to propose that we stop talking about social science and social philosophy, the science of economics and the philosophy of economics, political science and political philosophy, and speak henceforth about social science and meta-social science, economics and meta-economics, political science and meta-political science. Then the term "meta-science" can be identified, as we shall see later, with the term "philosophy" and possibly also with the term "humanities." But the distinction between science and meta-science will then become not merely that between a cognitive discipline and the systematic study of that discipline. It will also be the difference between a scientific discipline, using the ordinary methods and techniques of science, and a more sweeping, more critical, more analytic, and more speculative study of that very science, using whatever methods are appropriate to *its* type of investigation.

There is unfortunately a terminological matter here which is a recurring source of difficulty: the word "of" in certain contexts is ambiguous. Let us take the example of sociology. According to the terminology just proposed meta-sociology is primarily a study of sociology; it is a study of society only insofar as society is the basic concept examined by the sociologist. But note that the phrase "study of sociology" is ambiguous: it may mean that study which *is* sociology (this is probably the more common meaning), but it may also mean that study whose *subject matter* is sociology. Only in this latter sense is it a meta-science and can we attribute to it the use of the wide range of methods characteristic of the philosophical and humanistic enterprises. Although we might mislead if we say that the meta-sociologist studies the sociologist rather than society, there is a measure of truth in what we are asserting; for what the meta-sociologist studies is really the sociologist at work as a scientist, i.e., observing, describing, explaining, and formalizing, and also the sociologist at work on the fringe of his science, i.e., assuming the validity of certain methods, discovering unconscious presuppositions, and exploring the interrelationships between social values, moral values, religious values, economic values, and the rest.

Another confusion to be avoided by the proposed terminology is that provoked by the attempt to determine the "higher" or "more characteristically human" values in contrast to those which seem to play a less dominant role in the life of the individual. By general consent and historical tradition the three most significant goals are the "true, the good, and the beautiful." But here agreement stops. Is piety or spirituality, the goal of religion, also to be included in the list of specifically human values? One finds it difficult to locate either a time or a place when religious acts and feelings were not a signal part of the society. And the mere mention of the word "society" (and even the implications of the notion of moral good) suggests that a minimal social order is necessary for human survival. Why, then, should not favorable interpersonal relationships be integral to human living? And if social interrelations why not governments, and if governments why not trade and commerce? Indeed, since, as President Roosevelt pointed out, we must have not only freedoms "for" but freedoms "from"; freedom for health and at least *some* freedom from labor are necessary conditions for survival.

Since the question here raised has to do with the evaluation of values, which will be our concern in the last chapter, we shall attempt here only to establish that the line between the "higher" and the "lower" values simply cannot be drawn once and for all, but is established for an individual and for a culture, and is one of the most significant variables in human living. What we wish to clarify here is the usage of such terms as "science," "philosophy," "art" ("liberal" and "fine"), and "humanity." If we accept the proposal made earlier a science will be a first-level study, while a meta-science will be a second-level study of a first-level study. We have previously agreed that if the second-level study is simply a scientific study of science, as in the case of the psychology of science, we shall not call this scientific study a meta-science. But we hope to show later in the chapter that science itself is a mode of conscious, preferential behavior and therefore falls in the same general category as religion, art, morality, and the rest of the human valuational responses to the world. Now the sciences which study these modes of preferential behavior are the behavioral sciences, and if science is itself a mode of preferential behavior there must be a behavioral science whose task is to study it. Shall this be called a "science of science" or a "meta-science of science"? Fortunately the entire question turns out to be largely one of linguistic gymnastics. So far as I know there has never been a *general* behavioral science of science; there have been only such special sciences as we have seen — the economics of science, the political science of science, and similar studies. On the other hand, there has been and is a meta-science of science; it is precisely the philosophy of science and certain related studies. The explicit condition for this usage of terms is that the meta-sciences should not, in general, be restricted in their methods, attitudes, or approaches to those of the special sciences which they study. They may use the techniques of criticism, examination of presuppositions, and broadening of range of subject matter which are actually employed today in philosophy, the humanities, the arts, the "cultural" subjects, the normative disciplines, literature, humane letters, *belles-lettres,* and the rest of the "liberalizing" disciplines. What methods are used and should be used by these studies will not be examined in this book. I wish only to point out that my choice of the term "meta-science" to represent these second-level disciplines does not presume that they are sciences in the narrower sense in either method or content; they share with the sciences only the spirit of inquiry and the deep concern for truth. The summary of the material thus far examined in this chapter is given in Figure 11.

The transition from the lower stage of value enjoyment, through the stage of the scientific study of preferential behavior,

FIGURE 11

TABLE OF MODES OF PREFERENTIAL BEHAVIOR AND BEHAVIORAL STUDIES

I SUBJECT MATTERS (the "world")	II PURE SCIENCES (the sciences and critical metaphysics)	III META-SCIENCES (philosophy, art, humanities, etc.)
A. Conscious, Preferential Behavior and Goals	A. Pure Behavioral Sciences	A. Philosophies of Pure Behavioral Sciences
Political Behavior	Political Science	Philosophy of Political Science
Social Behavior	Sociology	Philosophy of Sociology
Work	Economics	Philosophy of Economics
Religion	Science and History of Religion	Philosophy of Religion
Morality	Descriptive Morals	Ethics
Art (creative and appreciative)	Description of Art Objects and Experiences	Aesthetics, Art Criticism
Health Behavior	Human Physiology and Psychology	Philosophy of Physiology and Psychology
Other typical forms of behavior, such as communicative, legal, teaching, learning, recreational, and so on.	Other typical behavioral sciences, such as linguistics, jurisprudence, pedagogy, psychology of learning, study of recreation, and so on.	Other meta-scientific studies, such as the philosophy of language, of jurisprudence, of pedagogy, of the psychology of learning, and so on.

to the "top" level at which the individual gains a new perspective over both his experiences and their scientific study is roughly concomitant with his biological growth and is, to this extent, not wholly under his control. Human growth, at least from later childhood to middle adulthood, is usually accompanied by an increasing dominance of the intellect, and value enjoyment involves a continually increasing participation of the rational faculties. Indeed, the plain truth is that, though this growth is more or less inevitable, we should not have it otherwise. Enhancement of value appreciation, both quantitatively and qualitatively, is one of the things in life which, for many of us, is basically desirable. We all reach, to a greater or lesser degree, the age of discretion—the age which is characterized by self-examination. Youth is the period of blundering enthusiasm. But maturity brings the sobering influences of principles, perspectives, and techniques; the adult must put away childish things. This does not demand the elimination of spontaneity and imagination, but it does require their chastening according to certain regulative principles. Values which in youth and adolescence were uncritically accepted, and merely felt and enjoyed, seem in adulthood to call for rational grounding. We want the enjoyment but we want also the knowledge of *what* and *why* we enjoy; we demand that the values have such objectivity as will make them persuasive toward other people; we insist that the goods of life be somehow or other fundamentally involved in the world in which we live. In a word, we want to *justify* our pursuit of values.

Growth in value appreciation, then, is often correlated with increase in rational justification. We seek to justify any enterprise when it is unsatisfying in the form in which it occurs and seems to call for further beliefs and convictions which will make it more properly rewarding.

4. Science as a Valuational Response

But what has become of science in this analysis? The answer should now be clear. Science as a form of inquiry is one of these basic forms of value pursuit.[17] It is, in all essentials, the same sort of thing as art, religion, morality, and the other valuational responses. One can devote himself to the pursuit of truth just as he can devote himself to the pursuit of beauty or God. Science is something which man does because he is convinced of the final worth of knowledge, and of its superiority to ignorance and error. Having discovered this, he selects means and instruments to its attainment. The choice and use of these tools are precisely what we mean by the scientific method, and the totality of activities and experiences enjoyed and lived through by the scientist in the employment of the tools and in the progressive realization of his

---

[17] R. B. Lindsay, *The Role of Science in Civilization* (New York: Harper & Row, Publishers, 1963), Chap. 3; and Herbert J. Muller, *Science and Criticism* (New Haven: Yale University Press, 1934), especially Chap. VIII.

goal are science itself. Thus science is a great enterprise of the human spirit which, like art, morality, and religion, has its history, including its crests of development and its troughs of decline and stagnation, its apostles and enemies, including both geniuses and charlatans on both sides, and its complicated cultural relations to the other phases of the human spirit.

But the statement that science is essentially like the other value pursuits should not be taken on faith; it demands proof. Such proof, of course, cannot be provided in any final sense, since resemblance is always a matter of degree. Clearly there are *some* respects in which science is like the other value pursuits, and the problem is to decide whether these resemblances are sufficiently important to warrant calling them to attention. The only test is to examine science in this light and see whether we gain new insight into its nature.

It must be recognized from the start that science, like the other forms of preferential behavior, exhibits itself primarily on the two lower stages. On these levels it is simply the scientific behavior of the scientist; it is the scientist at work, sometimes clearly aware of the methods which he is at the moment using and sometimes not. As we saw in the previous chapter, the fact that the scientist is not "thinking about his method" while employing it does not mean that he is ignorant of the nature of what he is doing, using his method badly, or incapable of turning his attention to it should the occasion arise. Presumably it is science on these lower levels which is most commonly intended when the word is used. Here the scientist is mainly using his method, but he can be thinking about it also; he is both acting and evaluating his actions concurrently; he is making wise choices among alternative acts with reference to the rules of logic and scientific methodology. Because of the cognitive nature of science we shall have difficulty deciding whether to characterize what he is doing as "science" or "meta-science"; as science it is a manifestation of a preferential activity in search of truth, just as art is a preferential activity in search of beauty; but as meta-science it is semicritical, at least partially aware of its methods and procedures, and to a degree conscious of its presuppositions. Consequently when we attempt to show the resemblances between science and the other forms of value pursuit, our approach must be meta-scientific; for we must examine these activities and goals analytically and critically.

When we inspect these resemblances we find them to be very striking indeed. Science *is* the pursuit of a value—a value which is very clear-cut and demanding. Whatever science may be, it is an unusually persistent effort to achieve knowledge, wisdom, and learning. We have already seen that the scientist is an individual who desires this value in some significant sense above all other values.[18] Whether he judges it to be in and of itself worth while, or merely desires it as an instrument by which humanity might be bettered or life made more comfortable is not relevant to the present issue. Man is both *homo sapiens*—a thinker, and *homo faber*—a doer. Evidently he may desire any one of the values of life as a means to the attainment of other values. But there must be at least one value which he prefers in and of itself. For the pure scientist the ultimate and final value is truth; for the practical scientist or the engineer this is merely an instrument to be employed in the realization of another value—say, better living—which then becomes an ultimate and final value. In either case science is the pursuit of truth.

Furthermore, the pursuit of truth would be without grounding unless man were dissatisfied with the superabundance of ignorance and error which he finds when he looks about him. Certainly there is no poverty of unsolved problems, mistaken information, and ignorance. The world, in fact, is a challenge because so many things we think we know about it just aren't so, and because so many of its aspects cry out for interpretation and explanation.

Motivated, therefore, both by his desire for truth and his dislike of ignorance, the scientist looks about for certain instrumental activities which he can employ in order to bring this ideal state somewhat closer to actuality. More simply stated, he begins to study the world for the purpose of learning as much as he can about it. Science then becomes all the methodological activities (observing, experimenting, measuring, idealizing, forming hypotheses and deductive systems) directed and controlled by the scientist, and all the psychological experiences (excitement and interest, as well as despair, discouragement, and frustration) through which he passes in his attempt to realize the goal of adequate knowledge. We are so accustomed to thinking of science in

---

[18] This does not mean, of course, as Mark Twain warned, that truth is so valuable that we should be economical in our use of it.

terms of its results that we tend to overlook the fact that science is, properly speaking, *scientists.* We have already had occasion to refer to the human aspects of the scientific pursuit. It is, like other value activities, a complex of behavior adjustments and emotional reactions which derive their unity from the fact that they are selected, employed, and appraised by the individual in terms of their efficacy in enabling him to achieve a state in which he can be said to have a satisfactory relation to the world about him.

Even more striking resemblances between science and the other forms of value activity appear as we penetrate beneath the surface. Characteristic among these is the feeling of "oughtness" which seems to be present in the conscious pursuit of all of the major values. This feeling appears whenever the conflict between the *desired* and the *desirable* becomes dominant. To say that a value is desired is, in a sense, to utter a truism; for if a thing is not desired it is simply not a value for the individual concerned. But to say that a value is *desirable* is not to assert a platitude; for if a thing is *desirable* it *ought to be desired,* and the plain fact of the matter is that many things which are desired by us are things which we feel ought not to be desired. To decide between what we want and what we feel we ought to have—this represents one of the most pervasive and unsettling conflicts with which man is confronted. In the moral sphere the opposition is between pleasure and duty, in the sphere of art it is between what we personally and privately like and what the informed critic tells us we should enjoy, in the sphere of religion it is between the many temptations which life offers and our responsibilities to God.

Science, too, exhibits this conflict. The presence of such decisions in science compels us to acknowledge that the scientist *is* concerned with values in some very profound sense. Whether he is conscious of the conflict between the "is" and the "ought" will depend on a number of factors—his early training, his moral stamina, and the general seriousness of his interest in science. Certainly he will have learned at some time that there are right and wrong ways of doing science, things which ought to be done and things which ought not to be done. Data ought to be collected, reported, and described with great care; theorizing ought to take its origin in facts and be subjected continually to the test of facts; results should be published as soon as they have been adequately verified; objectivity and impartiality should be the controlling ideals in the whole enterprise. The good—we can almost say "morally good"—scientist seldom has to make a decision of this kind since, like the good man in general, he more or less automatically chooses what is right. This does not mean, however, that value problems do not occur in science. The bad scientist, also, rarely has to make such a decision, for he always takes the easier road of slipshod methods, careless techniques, and unrestrained speculation. It means only that the competitive forces are so strong in science that the technician without scruples does not long survive; in science your sins will quickly find you out. Most scientists, therefore, having been adequately trained in scientific techniques, having developed the moral courage to hold to the principles of method, and having by word and act professed to a deep and abiding interest in science, habitually choose the approved rather than the disapproved techniques, and the fact that they are making a moral decision when they do this never enters their heads. The good scientist is a scholar, and a scholar knows both the ethics and the skills of his profession; the bad scientist knows neither.[19]

Another characteristic of value pursuits is their manifestation of the *qualitative* distinction between values—the distinction between higher and lower values. The classical illustration of this is to be found in the writings of John Stuart Mill. When the earlier utilitarians, attempting to construct an ethical theory, argued that we always act in any moral situation in such a way as to produce the maximum of pleasure for ourselves or for the social group, Mill replied that if this were true it would be better to be a contented pig than a discontented Socrates.[20] Pleasures differ, argued Mill, not only in quantity but in quality as well. Puzzling over problems, as exhibited in Socrates, may not produce *more* pleasure than wallowing in the mud but it produces a *better* pleasure. Whatever one may think of Mill's illustration, and even though one may argue convincingly that eating and sleeping *may* produce a higher quality pleasure than thinking, nevertheless qualitative distinctions between values *do* appear in all value pursuits.

---

[19] W. I. B. Beveridge, *The Art of Scientific Investigation* (New York: Random House, Inc., 1957), pp. 194–198.

[20] J. S. Mill, *Utilitarianism, Liberty and Representative Government* (London: J. M. Dent & Sons, 1910), p. 9.

In science the distinction is illustrated by the difference between particular truths and general truths. If we were to say that the mass of a particular piece of iron is 17.2 grams, or that a certain object is an example of *canis familiaris* we should be uttering descriptive truths, while if we were to mention Boyle's law or the evolutionary theory we would be affirming general or theoretical truths. Both are truths and therefore fit into the general picture of science. But the latter seem to be on a higher level, and to be more significantly involved in the forward movement of science.[21] Thus there arises, at least in the minds of some people, a distinction between the aristocrats of science who concern themselves with the "higher" truths—laws, basic presuppositions, integrating theories, and wide generalization—and the plebeians of science who, at least in the eyes of the aristocrats, are mere fact-gatherers and "gadgeteers."[22] Whether the aristocrats and the plebeians have been properly identified here is not pertinent to the discussion. One might argue that since all theory must rest on fact, and there are clearly times when one fact is worth a hundred theories, those who are concerned with particular facts are the real aristocrats of science, while the theoreticians are scientific philanderers and wastrels. The point is not whether particular facts are high quality truths or low quality, but whether the distinction does apply in some sense to scientific truths. General truths are not more true than specific truths, nor is the reverse the case; both are truths, just as ping-pong and poetry are pleasures, yet one seems preferable to the other on grounds of quality. That science does exhibit this distinction strengthens the reasons for a closer identification of science with the other value pursuits.

We have attempted to show in this chapter that science has much in common with the other basic forms of value pursuit. In order to do this we called attention to five features of science. (*a*) It is the pursuit of a highly esteemed value—truth. (*b*) It receives its drive largely from a realization that there is too much error and ignorance in the world. (*c*) It sets up activities designed to help in the attaining of its goal. (*d*) It exhibits the conflict between the *is* and the *ought* which is common to all forms of value pursuit. (*e*) It exhibits also the distinction between *higher* and

[21] I. B. Cohen, *Science, Servant of Man* (Boston: Little, Brown & Co., 1948), p. 55.

[22] Beveridge, pp. 198–199. See also above, p. 164.

lower values which is characteristic of the other modes of valuational response.

But the demonstration of this truth, supposing it to have been successful, does not commit us to the complete union of science with any of the other modes of preferential behavior. Science is still, in its most significant respects, a unique activity of the human spirit. It is certainly not to be *identified* with art, with religion, or with morality. It has its own values and its own methods. Part of its uniqueness lies in the special ways in which it interacts with the other forms of valuational response. The ensuing chapters will deal with this specific problem.

Through Science to Philosophy

Little has been said thus far about philosophy. In the opening pages of our discussion we suggested that science, philosophy, poetry, and religion may have had a common ancestor in the wonderment and fear which possessed man in the presence of a hostile nature. Gradually, in the course of his intellectual, spiritual, and cultural development he separated these forms of response according to the specific value which was involved in each case. Poetry, being a form of art, seemed to be concerned primarily with beauty; religion proved to be an attempt to develop, by a complex of attitudes and acts involving worship and sacrifice, a feeling of security in the face of a terrifying world; science and philosophy, by sharp contrast, seemed to be interested only in truth.

But though little has been said about philosophy, most of what has been done thus far in our study *has been* philosophy. As we saw in the previous chapter the description and analysis of the scientific method, when carried on as a critical study of the pursuit of truth and its presuppositions, is an important part of philosophy. We are now ready to turn to the direct examination of the nature of philosophy and some of the relations, simple and complex, which it bears to science.[1]

The general conclusion of the last chapter affords a good starting-point for our discussion. There it was suggested that human living breaks up into a number of forms of valuational response. We listed as the kinds of such response: inquiry, art, religion, morality, social behavior, political behavior, work, and play. There was no suggestion that this catalogue was meant to be exhaustive. Philosophy and science, since they are both concerned with truth, are clearly kinds of inquiry. Being modes of inquiry each exhibits, as we saw in our first chapter, five aspects. It has a subject matter, something about which it talks and endeavors to build up knowledge. It is a method of study, a type of procedure by which given facts are interrelated, interpreted, and explained. It is carried on by human beings who control and direct the enterprise toward the maximum acquisition of knowledge. It terminates in a body of truths, descriptive, explanatory, and formal, about the world. Finally, it is motivated by the desires and interests of the investigator, be they the theoretical interests in knowledge as such or the practical interests in improving human living.

But how are science and philosophy differentiated? Unfortunately neither etymology nor history helps us to answer this question. Etymologically, "philosophy" means love of wisdom, and "science" means knowledge; unless, therefore, we have some way of distinguishing between wisdom and knowledge we have learned nothing. Historically, until well into the nineteenth century, philosophy and science were one and the same thing. Aristotle was clearly both philosopher and scientist, and we can separate these two parts of his learning only on the basis of what we now include under each of the terms, not according to what was for him a sharp distinction. The development of thought from the days of the Greeks down to the present has been characterized by a progressive breaking up of the broad fields of philosophy into more and more limited areas, and the word "science" has become

[1] R. B. Lindsay, *The Role of Science in Civilization* (New York: Harper & Row, Publishers, 1963), Chap. IV; C. J. Ducasse, *Philosophy as a Science* (New York: Oskar Piest, 1941); and Herbert Dingle, *Through Science to Philosophy* (Oxford: Clarendon Press, 1937), especially Chap. I.

increasingly applied to the various studies in these more restricted fields.

The breaking up of philosophy into more specialized fields—natural philosophy which became physics, and mental philosophy which is now psychology—has produced some confusion in usage. For example, the American Philosophical Association is the official organization of philosophers in the United States; but there is also an American Philosophical Society which is an organization of scientists, founded, incidentally, by Benjamin Franklin and hence employing the terminology which was current in his day. The attempted breaking away of science from its philosophical ancestors has led many writers to search for a principle according to which science and philosophy can be differentiated. Some have felt that philosophy began as a rather vague and confused study, closely bound up with man's hopes and fears, but developed into a legitimate enterprise when it began to insist on accuracy and precision, even mathematical precision, in its concepts and methods. Once this had been achieved it was entitled to break the unfortunate associations of its past by calling itself "science" rather than "philosophy." Others have felt that philosophy, even when it became precise, was still speculative, and that it could be called "scientific" only when it replaced hunches and guesses by probable knowledge, and probable knowledge by certainty. Still others, looking into philosophy, saw that certain of its parts made no progress whatsoever but continued to be a rehashing of old problems, while other parts made discoveries, definitely refuted old theories, and advanced both in scope and certainty. They suggested, therefore, that philosophy could be divided into two parts—a fertile and productive part which should be called "science," and a barren and unproductive part which should continue to be called "philosophy."

While there is a certain soundness in these attempts to distinguish science from philosophy, they do not go to the heart of the problem. If there is a fundamental difference between the two types of study we may reasonably suppose that it lies either in method or in subject matter. And if the philosophical method has been perfected through the ages, any vagueness in its results, any uncertainty in its conclusions, and any lack of progress in its development would seem to be due to what is talked about. As we have frequently seen, in science itself there are some facts which are directly accessible to observation; these we can talk about with a fair measure of success. But there are other facts whose existence seems undeniable yet which cannot be observed; these must be discussed in vague and conjectural terms, and progress in understanding them is slow indeed. Is it not possible, therefore, that there is actually a wide variation among facts—in the clearness with which they are given, in their availability for observation at particular times and places, in their accessibility to the use of recording devices and other instrumental aids, in their capacity to be classified, associated, correlated, and measured, and in many other ways? Perhaps, then, the difference between science and philosophy lies in the kinds of facts which each selects for study. The two are not rivals, but each has its field marked in advance. If we grant this, three possible distinctions immediately suggest themselves. (1) Science studies specific facts; philosophy studies general facts. (2) Science studies parts of the universe; philosophy studies the universe as a whole. (3) Science studies nature and our conscious, preferential behavior; philosophy studies our basic valuational behavior and the presuppositions which underlie it.

1. Specific Facts vs. General Facts

It is no accident that we usually talk about the "special" sciences; strictly speaking there is no science in general—there are only the sciences. What distinguishes one science from another is its subject matter. Logic and mathematics talk about such abstract entities as elements, sets, numbers, orders, and abstract relations indicated by such words as *implies*, *or*, *not*, and *and*. The physical sciences inquire into mass-energy, force, motion, and allied concepts. The biological sciences investigate life. And the behavioral sciences study our preferential responses. Now it is obvious that the way in which these subject matters differ from one another is in degree of generality. Life is more general than conscious, preferential responses because all valuational beings are alive, but not all life involves free choice. Mass-energy is more general than life because all life is mass-energy, but not all mass-energy is alive. Abstract structure is more general than mass-energy because although all mass-energy is capable of abstract structure, logical form is also applicable to "immaterial" entities.

As a consequence these four kinds of fact can be arranged in a column which indicates not only that abstract structure is the most general subject matter and preferential behavior the most special, but that the more special depends on the more general, with a corresponding dependence among the sciences. The "higher" sciences rest upon the "lower." This is shown in Figure 12.

FIGURE 12

TABLE OF SUBJECT MATTERS AND SCIENCES

I SUBJECT MATTERS (the "world")	II PURE SCIENCES
A. Conscious, Preferential Behavior	A. Pure Behavioral Sciences
B. Life	B. Pure Biological Sciences
C. Mass-Energy	C. Pure Physical Sciences
D. Abstract Structures	D. Logico-Mathematics

The construction of such a table as this gives rise to an interesting conjecture: Is there, perhaps, a study which has a subject matter which is even more general than that of mathematics? Is it possible to find a kind of inquiry which is so wide in its scope that its range extends over the entire field of "objects," actual and possible? Such a study, if it could be found, would presumably be the most basic of all the sciences and would be that science on which all others ultimately rest.

Tradition has applied the name "philosophy," in the sense of "critical metaphysics" to this inquiry.[2] Its subject matter is the most general "entities" (there is no other word in the English language suitable to characterize them)—entities which we find not only in our actual universe but would find in every fanciful and thinkable universe. Try, for example, to imagine a world in which there are no *things*, no *qualities*, and no *relations*. We have no great difficulty in picturing a universe in which things have qualities and relations quite different from those which they possess in our world; in myths, fairy tales, and dreams, we have experiences of such imaginative realms. But if we try to represent

[2] C. D. Broad, *Scientific Thought* (New York: Harcourt, Brace & Company, Inc., 1923), Introduction.

a world in which there are no things whatsoever — in talking about such a world we could use no nouns, pronouns, proper names, or other substantival terms — we find it literally impossible. Even supposing we allow our world to have things but no qualities or relations, we are no better off; for we could have no way of identifying our things or connecting them with one another. We have similar difficulties in conceiving of a world in which there is no time or in which there is no space. To speak of mathematics and logic as dealing with timeless and spaceless worlds is permissible, as we have already seen, but these worlds take on meaning only as abstractions from our actual world; without a world of time the word "timeless" would be meaningless. Such terms as these — commonly called categories: *thing, quality, relation, time, space, actuality, possibility, necessity* — are so general in scope that they include everything mentionable; they extend over the entire universe and nothing could be excluded from them. Definition of such general terms is, of course, impossible, since definition always proceeds towards terms of greater and greater generality. But if the terms which we are trying to define are the most general notions in our entire vocabulary they are *sui generis*, i.e., they are their own *genera*, hence unique. The best we can do is to "talk about" the terms in question. Most of us know in a vague sort of way what they mean, and the philosophical task is to render them as clear as the difficulties of the subject matter permit.

An addition to our table of the sciences is therefore called for. This is given in Figure 13, and shows the position of critical metaphysics with reference to the special sciences.

Because the order of the subject matters, passing from *A* to *E*, is based on increased generality, and philosophy, in this one of its aspects, is identified with the study of what is most general in the world, the sciences, passing from *A* to *E*, will reflect the same increase in philosophical character. The behavioral sciences, including all valuational studies, are the least philosophical because they deal with what is highly specific in the universe — conscious, living matter. The biological sciences are more philosophical because they include conscious beings in their subject matter, but discuss all life. The physical sciences are still more philosophical because they talk about all mass-energy, which includes both the living and the non-living. The logico-mathematical sciences are commonly recognized to be highly philosophical in character;

Plato insisted that those who were to study philosophy under his tutelage first acquire a knowledge of mathematics. Critical metaphysics is, in this sense of the term "philosophy," the most basic of the sciences and therefore the most philosophical.

FIGURE 13

TABLE OF SUBJECT MATTERS, SCIENCES, AND META-SCIENCES

I SUBJECT MATTERS (the "world")	II PURE SCIENCES (including metaphysics)	III META-SCIENCES (philosophy, arts, humanities)
A. Conscious, Preferential Behavior	A. Pure Behavioral Sciences	A. Philosophy of Behavioral Sciences
B. Life	B. Pure Biological Science	B. Philosophy of Biological Science
C. Mass-Energy	C. Pure Physical Science	C. Philosophy of Physical Science
D. Abstract Structures	D. Logico-Mathematics	D. Philosophy of Logic and Mathematics
E. The Categories	E. Critical Metaphysics	E ....

2. PARTS VS. WHOLES

But the special sciences are "special" in another sense: the subject matter of one science differs from that of another. When we say that a scientist is a specialist we mean just this; he is concerned with plants and not with animals, or with the heavens and not with man's morals, or with the structure of the earth and not with trade and commerce. Today, of course, specialization has gone much further than this. A physicist is no longer a physicist but an atomic physicist, or a metallurgist, or a specialist in cosmic rays. A biologist is a geneticist, or a pathologist, or an embryologist. The day of the old-fashioned medical practitioner has gone and we now have pediatricians, ophthalmologists, gynecologists, and others too numerous to mention. The world has become so complicated and the knowledge of the regions within it so extensive that no man is able to know all things, or even all that is involved in one science. Division of labor is now an accomplished fact. Greater efficiency is thereby achieved in any given field, for a wide range of facts may be neglected by an investigator in order that he may concentrate on the facts which are relevant. The physicist can carry on his own work effectively without knowing sociology; the astronomer can do a good job without knowing biology.

Such specialization, however, leaves a very important task undone. The world does not break itself up into these parts; only the mind of man can do this. There remains, consequently, a very important and very difficult problem: How can the results of the various sciences be put together into a *universe*? We may be wrong, of course, in supposing that they can be combined into a consistent and unified whole; perhaps we live not in a *universe* but in a *pluriverse*. But even if this is the case, even if our world is a very loose-knit aggregate, nevertheless *some* sort of relations must hold among the parts, and we should try to discover what these relations are. What is the relation between mind and matter, between the stars and society, between life and the earth, between nuclear fission and human happiness? Once we have discovered these relations our job is to put the sciences together, in a picture-puzzle fashion, so that we may see what the resulting whole is. The task is a difficult one and none of the special scientists considers it his job; because of the magnitude of his own special assignments he does not have either time or energy for it. But the problem is a genuine one and, as recent events have shown, an urgent one. Tradition has also assigned this problem to that part of philosophy which is sometimes called "speculative metaphysics." So conceived, philosophy is a search for a unified picture of all things, for what the Germans call a *Weltanschauung* —an intuition of "things altogether." This is philosophy in its integrating and unifying aspect. Matthew Arnold said of Sophocles, "He saw life steadily, and he saw it whole." The job of philosophy is to see the world steadily and as a whole.

Here, again, we can see that the distinction between science and philosophy is one of degree rather than of kind. Many sciences, realizing the possible errors in isolating objects from the backgrounds in which they occur, have tried to adopt a synoptic point of view and to introduce on a certain level a degree of integration. Geography is, according to this conception, one of the most philosophical of the sciences, for it attempts to unify all that exists and goes on in a certain locality on the surface of the earth —its social and its cultural life, its animal and plant life, its climatic and geological changes, and so on. The area-language studies which are commonly taught in universities today as preparation for membership in the Peace Corps, and which were an important part of the Army Specialized Training Program of World War II are similar attempts along these lines. History is another of the philosophical sciences, since, even if it restricts itself to social and political events, its broad temporal sweep requires a high degree of organization and integration. But the task of interrelating all the sciences in the most comprehensive way possible is reserved for philosophy.

One special part of the integrative job is sufficiently important to deserve mention here, though I shall postpone its detailed examination to the last chapter. It is concerned with man's attempt to fit the many values of life into a consistent pattern—a pattern which will enable him to make rational choices when he is confronted with conflicts among the various goals. This is the problem of setting up a plan for living.

3. Nature and Behavior vs. Critical Evaluation

The last of the relations between science and philosophy requires no detailed consideration at this point, since it was developed in the previous chapter. The fact that inquiry is a value pursuit has been abundantly shown; science, art, religion, social behavior, and the rest of the preferential activities and values are deemed necessary, at least by many people, for the full living of the individual life. Suppose we temporarily classify all of the valuational pursuits into two classes: the "intellectual" and the "non-intellectual." The principle of this twofold division is whether or not the *ultimate* goal of the activity is or is not *truth*. If proper interpretation is given to the italicized words, only inquiry will fall in the first class and all others will fall into the second. For example, art is a search for beauty, not truth; religion is a search for spirituality, not truth; political behavior is a search for good government, not truth; and so on with the other preferential activities. Inquiry, alone, will be a search for truth.

But we pointed out that all of these "non-intellectual" pursuits, as the individual becomes more mature, tend to *call upon* inquiry as a tool for rendering choices in all of these areas more satisfying. A more satisfying experience arises to the extent to which the pursuit of the values becomes more conscious and the techniques for attaining them more selective. The distinction between effective and ineffective methods is replaced by the distinction between right and wrong methods, and truth begins to emerge as a definable and deliberately sought-after goal. Here inquiry is functioning as a behavioral science, and the obvious principle involved is that in general one's social, political, moral, religious, and other modes of value response become more satisfying when based upon studies of sociology, political science, descriptive morals, and comparative religion.

However, as still further demands for increased satisfaction appear in all of the value areas, inquiry enters once more. At this stage it is meta-science or one of the humanities. It is the criticism of one's own criticism. The individual wants to be as certain as possible that he understands his goal and that he is fully aware of the presuppositions which would justify his pursuit, and of the risks of failure which he must run in order to achieve his goal. In uncovering presuppositions we must ask ourselves such questions as the following: What assumptions do we make when we set up goals and try to achieve them? What must be true about ourselves and about the world if these goals are to be attained? *Can* the goals be attained, or are we, perhaps, engaged in a hopeless activity? But after we have determined what the assumptions of a given value pursuit are, we have still another question: are the presuppositions true? If they are—that is, if they are such as can be reasonably believed—the value experience has received its proper justification; if they are not true, or cannot be shown to be justifiable in any way, the value pursuit is without justification and must remain an experience which is tainted with frustration and which may end in disappointment and unhappiness.

Inquiry, therefore, plays a threefold role in human living. In the first place, it is simply one of the many value activities, all substantially alike, pursued by individuals who usually, as they grow to maturity, seek a rational justification for the pursuits they are carrying on. In this role inquiry is simply science, and it demands justification in essentially the same way that the other value pursuits do. In the second place, inquiry has the task of "intellectualizing" all of the value pursuits by providing scientific studies of the activities themselves; in this role inquiry is substantially identified with behavioral science. But in its role as meta-science or one of the humanities inquiry has still a third task—to uncover the presuppositions of all of the value pursuits and to examine their possible foundation in fact. This requires a much more critical analysis than can be carried at the level of the behavioral sciences, and it should therefore be properly designated as "meta-science" or a "philosophy" for each of the value pursuits studied. But the fact that this critical investigation is of *all* of the value activities, and the fact that inquiry is one of the value activities, leads to the existence of a unique study which might well be called "meta-inquiry." In this study the goal which is pursued can be achieved only by using that goal itself. In morality we seek the good and the truth about the good; in religion we seek spirituality and the truth about spirituality; in political behavior we seek good government and the truth about good government; only in meta-inquiry do we seek the truth and the truth about truth.

In conclusion, reference should be made to a strong movement in recent philosophy, commonly called "linguistic analysis." It had its origin in the position of the logical positivists, whose views were briefly discussed earlier,[3] but its perspective has broadened to the point which considers philosophy to be *definable* as the analysis of language. I should like to indicate that the view of philosophy which has been developed in this chapter is completely in accord with this interpretation of philosophy, provided it is not unduly restricted. The linguistic analysts tend to stress the fact that the best way to identify a value experience is by the language commonly used by one who is having the experience. For example an individual who is making moral decisions frequently uses such words as *right, wrong, duty, conscience, ought, desirable,* and many others. And a person who is behaving religiously often mentions *prayer, faith, God, sin, providence, worship, repentance,* and *hope.* The same is true of the other value experiences. Now it seems obvious that the presence of a value activity is not *exhaustively* described by the words which the participant in the experiences uses, because he may perform certain moral or religious acts without saying anything, and he probably has certain moral or religious feelings and attitudes which he communicates to no one. Again the same is true of the other value pursuits. Thus if we were to restrict our knowledge of these prefer-

[3] See above, p. 86.

ential activities to the words used by those engaged in value pursuits, we should have only a partial indication of the total contents of the experiences.

But if we move now to the level of the behavioral *sciences*, which *study* these modes of behavior, the same words will recur and many others will be added because in the scientific method we can extend the range of given subject matter from the words which are given as the starting point, to hypotheses and theories which we discover or improvise in order to explain why these words are used. Furthermore (and this is true only in the behavioral sciences), we can ask the participants to tell us about their feelings and attitudes. From both these sources we gain new data—data which are, to be sure, still words, but words which are so numerous as to give us an adequate representation of the experiences which we are studying.

Now, finally, if we recall that for every science there is a meta-science, and if we think of a science as a system of symbols (words) referring to a certain subject matter, then a meta-science will be a meta-language[4] which describes and explains an object language which in turn describes and explains a mode of valuational behavior. This permits us to use the word "philosophy," insofar as it is identified with meta-science, as the analysis of language. All that this actually means is that instead of defining the "philosophy of religion" as the study of prayer, worship, God, and sin, we define it as the study of the words *prayer*, *worship*, *God*, and *sin* as these combine into something which might be called the *language of religion*. Similarly "ethics" becomes the study of the language of moral behavior, "aesthetics" becomes the study of the language of art; "political philosophy" becomes a study of the language of politics; and the "philosophy of science" becomes a study of the language of science. Thus Chapter 3, and especially the summarizing diagram at the end of the chapter, become important aspects of our study of science. This new emphasis on language is clearly indicated by the titles of many books on valuational behavior which have appeared in recent years.[5] There

---

[4] See above, pp. 59–60.

[5] W. Elton, *Aesthetics and Language* (Oxford: Oxford University Press, 1953); G. A. Carver, *Aesthetics and the Problem of Meaning* (New Haven: Yale University Press, 1952); R. M. Hare, *The Language of Morals* (Oxford: Oxford University Press, 1952); C. L. Stevenson, *Ethics and Language* (New Haven: Yale University Press, 1944); H. D. Lasswell, *The Language of Politics* (New York: George W. Stewart, Publisher, Inc.

is no harm in this linguistic approach provided one realizes that by emphasizing the *language* of the valuational behavior he does not overlook the behavior itself or the possible discrepancies between the content of the behavior and the language by means of which this behavior is expressed.

1949); *New Essays in Philosophical Theology*, ed. A. Flew and A. MacIntyre (London: SCM Press, 1955); and Paul Taylor, *Normative Discourse* (New York: Prentice-Hall, Inc., 1961).

Religion in an Age of Science

One of the oldest conflicts in the history of thought is that between science and religion.[1] The range and variety of the solutions which have been offered to this problem are a tribute both to the seriousness of the problem and to the intellectual skill of mankind. Religion has always been for those who participate in it a vital matter—an affair not merely of the understanding but of the heart as well. Frequently religion has been adopted as a last resort, when all other values in life have been found incapable of providing the desired comfort and happiness. Thus it has been precious, so much the more so because if it were to fail then life itself would fail. As a consequence the struggle between religion and science in the history of thought has been long and bitter.

Just why this particular conflict has been singled out for so much attention in the history of thought will be explained in a moment. We have already seen that human life is full of conflicts,

[1] A. D. White, *History of the Warfare Between Science and Theology* (New York: D. Appleton & Co., 1914), and E. A. Burtt, *Metaphysical Foundations of Modern Physical Science* (New York: Harcourt, Brace & Company, 1925).

even for people who seem happy and well adjusted. Life is so constituted that we cannot have all the things we want. Sometimes the difficulty lies in the values themselves; they are mutually exclusive and cannot coexist. Sometimes the difficulty lies in ourselves; we do not live long enough, or we do not have the health and energy required, or we do not have the skills and capacities which are needed. But in none of these respects is the conflict between science and religion unique. Truth and beauty often conflict: the most beautiful portrait is frequently not that which reproduces photographically the likeness of the model. Religion and art sometimes conflict: some of the old-time hymns and revival songs, though they are frequently examples of inferior music, have been effective as ritualistic devices for arousing religious fervor and strengthening faith. Truth and morality are often at odds: the spy who is caught by the enemy is not morally obligated to tell the truth, the doctor at the beside of a dying patient need not tell him his true condition, and only the most boorish man will express an honest opinion when he sees milady in the newest fashion creation. So the conflict runs through the gamut of human value experiences.

1. Religion and Science

It will be impossible because of the limitations of space to present a detailed examination of all of the major value conflicts which characterize human living. Something has already been said of the moral aspects of science in Chapters 8 and 9, and attention will be turned to another aspect of this problem in Chapter 14. The problem of the interrelations between scientific values and educational values will be examined in the next chapter. Chapter 13 will be devoted to the role of science in social behavior, and some reference will be made at this point to the relations between science and government.

Our selection of the conflict between science and religion as a problem for detailed analysis lies in the fact that religion alone among the non-cognitive value experiences includes as an important component a theory about the natural world, its origin and destiny, and the place of values in it. Since the natural world, with or without the values which it may contain, is the essential subject matter of science, a conflict is almost certain to arise between the view of the world which is demanded by the religious experience and that which science uncovers. This has been particularly true in the history of astronomy. Its view of the place of the earth (and consequently of man on the earth) in the rest of the universe has been of paramount consideration in upsetting the fervently held views of many religious people. Biology has also come in for its share of criticism, for the evolutionary theory seemed to make man continuous with the animals and thus to destroy his sharp superiority over them in possessing an immortal soul.

Furthermore, in the comparison of religious and scientific values, and in the contrast between these and all other values, we can see the importance of the question of the *objective status* of values. Clearly for science there can be only one truth, and this must somehow be embodied in the universe. Like the other values it is the object of human desire; but unlike these values it cannot reside wholly in the desires. Beauty can be subjective, but scientific truth cannot lie simply in the opinion of the scientist. Religious values, for the religious man, are more like truth values than beauty values: they must be objective if the religious experience is to be a satisfying one. In this respect religious values are more like moral values than the other goals of life, and this accounts for the strong moral component in most religions.

Probably both the long history and the great intensity of the conflict between science and religion are to be explained partly by the fact that the issue cannot be sharply focused. "Religion" is one of the hardest words in our entire vocabulary to define. We cannot proceed as we did in the case of "science" by collecting rough examples of the term, and then try to extract certain common properties. Religion has existed in so many times and places, under such a variety of forms, and in such intimate relations to the many biological, social, and psychological aspects of man's life that we are unable without great difficulty to extract the highest common denominator from all its manifestations. We cannot, in other words, start with all the kinds of experience which are called "religion" and expect to abstract a definition which is adequate to the entire range of such cases. The only alternative is to begin with a definition which is more or less arbitrary, even though this involves throwing out many instances on the grounds that they are not truly religions. Following this pattern many scholars have defined religion in such a way as to make a conflict with science impossible. A large number of scientists in recent years—men of unquestioned integrity and of strong religious convictions—have chosen this solution to the problem. Others have felt that while there was a conflict between religion and the old, materialistic science, the new science, with its idealistic leanings and its recognition of gaps in the causal rigidity of nature, is thoroughly compatible with religion and, in fact, contributes definitely to its support.

The approach which we have made thus far in our study suggests a way of attacking the problem. Let us return to our consideration of science as a mode of inquiry. We concluded that science exhibited five features.[2] It is (1) a pursuit of truth which (2) originates in the unpleasant fact of ignorance and error, (3) employs certain activities and methods in its support, (4) gives rise to feelings of "oughtness" in the scientist, and (5) exhibits distinctions between "higher" and "lower" values. Now let us look at religion.

The goal of religion is a way of life, an attitude toward the world which has been variously called "spirituality," "holiness," and "piety." The confidence that the goal can be attained, at least by approximation, is called "faith." The negative value is the uncertainty, insecurity, and fear which arise in man when he finds himself confronted by a world which is more powerful than he is and apparently unconcerned about his happiness or survival. If he is a member of a primitive society living close to nature he has adequate cause for terror in earthquake and hurricane, in starvation and attack by wild animals and enemies, and in disease and death. If he lives in modern society he is protected somewhat from the ravages of nature but he has the constant threat of war, accidental death, loss of fortune, unemployment, and all that goes with a highly integrated social order. The realization of this constant danger leads him, as in the case of science, to imagine an ideal world which would provide him with hope, peace, comfort, and security. Such a realm could be actualized only by changing his present environment (since this is certainly not the kind of world which *does produce* these feelings), or by changing himself (since he does not actually *have* these feelings),

---

[2] See above, pp. 187–193.

or possibly by changing both. He then sets about to realize this ideal, so far as this is possible, through attempts to control and modify nature by propitiatory acts and prayers; through efforts to bring about a change in the social world by persuasion, teaching, and leading an exemplary life; and through activities such as meditation, worship, and devotion, which are designed to transform his own personality. The vision of the ideal affords him comfort in times of sorrow and adversity, provides him with something before which he may practice humility, and inspires him to continue the struggle against the forces of evil and destruction.

The feeling of "oughtness" is a marked characteristic of religion. When religion is closely bound up with an ethical code it defines certain *vices*, and these, of course, ought not to exist. But a religion may also define certain *sins*, and these ought not to exist because they violate the divine law and prevent the religious man from bringing about those changes in the world and in the man himself which alone will enable him to develop an attitude of piety. Here the contrast between the *desired*, which is usually productive of pleasure, and the *desirable*, which commonly involves self-denial, is sharp. Here also the qualitative aspects of values become apparent. While the pleasures of eating, drinking, and merriment are not excluded from life, at least in the more liberal Christian tradition, they are clearly believed to be inferior in quality to the satisfactions of prayer, meditation, worship, devotion, and proselytizing.

In accepting this parallel analysis of science and religion we should recognize what assumptions have crept into our argument. Science has become generalized. We have left the area of physics, or chemistry, or biology, and are talking about science simply as a mode of inquiry. In this sense science is identified with *investigation, scholarship,* and *the pursuit of knowledge.* It includes philosophy in the three roles mentioned in Chapter 10. The scientist is curious about the world and, as a result of this attitude, engages in certain activities which one who does not have this outlook refrains from doing because he is not interested in them, because he considers them unimportant, or even because he judges them to have harmful results. Similarly, in talking about religion we have risen above Presbyterianism and Methodism, above Protestantism and Catholicism, and above Christianity, Buddhism, and Confucianism. The religious man is simply one who experiences certain feelings and needs when he is confronted by the world, and as a result engages in certain activities which one who does not have these feelings avoids because he does not have the motivation, because he considers the activities unnecessary, or even because he judges them to have bad consequences for man and the world. The "conflict" between science and religion, therefore, is one of value perspectives. Does man "adjust" to the world better by understanding it (through science), or by fortifying himself against its evils and strengthening his will to improve it by developing a reverential attitude toward it (through religion)? Does one live by knowledge or by faith? When the question is formulated in this way it cannot be answered *in* religion or *in* science, for it concerns a choice between two basic values of life. We shall attempt to face this question in our last chapter.

2. The Presuppositions of Science

Here we should like to suggest another approach to the problem—an approach which is based on the conclusions of Chapters 9 and 10. The reader will recall that in the consideration of the relation between science and philosophy we discovered that philosophy in the sense of meta-science, seeks the rational groundings not only for science but for all of man's value activities. This demand for justification seems to be present simply because the value activities prove less satisfying as we reach adulthood. We begin to have misgivings as to the legitimacy of the pursuits, as to the attainability of the goals, and as to the reasonableness of making the required sacrifices. We adopt a more inquiring attitude and demand rational justification for what we are doing. Philosophy aids this effort in two ways: It helps us to uncover the presuppositions of our activity; it tells us what assumptions we are making in carrying on the enterprise, and what would have to be true if the goal is to be attainable. Furthermore, it assists us in evaluating these presuppositions, i.e., in deciding questions of truth and falsity with regard to them—whether truth and falsity really apply to them and, if so, in terms of what evidence; whether they are actually true or false in any given case; and so on. If the presuppositions are such as can be believed, the value pursuit receives its proper justification; if they are not, it is without justification and we must remain satisfied with a less rewarding experience.

Let us see what is brought to light when we call in philosophy to examine the presuppositions of science. The pursuit of truth seems to involve at least the two assumptions of all value experiences. First, truth is one of the supreme values in life. Acceptance of this statement requires an understanding of what is meant by truth, what is meant by a supreme value, and how conflicts between values are to be resolved. Second, truth is attainable, at least by approximation. Acceptance of this statement requires making explicit the conditions under which it could conceivably be realized. This necessitates listing such factors as the rationality of nature, its objectivity and uniformity, its operation according to causal laws, the validity of the laws of logic, and the adequacy of communication. These are simply very concise ways of stating that we cannot understand the world unless it is knowable, unless it is in some sense independent of our knowing activities and maintains a certain regularity in its operation, unless the rules of induction and deduction which we employ are valid, and unless we are able to work cooperatively with one another and communicate the results of our studies. *Unless* these and other assumptions about truth are such as can be believed, we had better give up the whole enterprise; on the other hand, *if* they are such as can be believed, science has received its proper justification and will continue to prove a rewarding and satisfying activity.

But can philosophy meet these demands? Can it show that truth *is* worth while, and that nature *has* the required objectivity, rationality, and uniformity? The assignment does not look too promising, since we seem to be required to demonstrate certain *truths* about *truth*. Let us explore the possibilities. There are only three.

3. Justification of these Presuppositions

The first possibility is that the assumptions which are required to justify science can themselves be established by the very techniques of science. Certain writers, apparently, believe that this can be done. For example, C. E. Ayres states what he calls "the scientist's creed":

> I believe in atoms, molecules, and electrons, matter of heaven and earth, and electrical energy its only form. I believe in Modern Science, conceived by Copernicus, borne out by Newton, which suffered under

the inquisition, was persecuted and anathematized, but rose to the right hand of civilization as a consequence of the fact that it rules the quick and the dead. I believe in the National Research Council, the communion of the scientists, the publication of discoveries, the control of nature and progress everlasting. Amen.[3]

Now the most interesting comment on such a group of statements is that it contains almost nothing of what would ordinarily be found in a creed. The term "creed," though commonly restricted to religion, is applicable to all the value pursuits, and refers simply to the presuppositions which are required to justify the preferential activities involved. But most of the statements in this creed are not of this kind; they are hypotheses which, if we assume the methods of science, can be properly verified and established. One would not ordinarily say that belief in molecules is a presupposition of science, or that belief in the existence of energy is part of the creed of science. These are simply theories which are highly probable because they have been confirmed by the scientific method. Even the historical statements which this so-called "creed" contains could be established in this way.

If we take greater care in the formulation of the creed of science the task of proving the statements which it contains runs into difficulties. Especially is this true if we try to prove the assumptions by using the very method of science. Let us suppose, for example, that the principle of the uniformity of nature—that the world is guided and regulated by universal and necessary laws—is simply a theory like the molecular theory, which must be confirmed in terms of its predicted consequences. If we find that its implications are everywhere supported by the facts, and nowhere disconfirmed, we are entitled to say that it is very highly probable. But this will not do. We cannot prove the principle of the uniformity of nature by using the scientific method, for this method is justified only if the principle is true. If nature is not uniform we have no business employing the scientific method, for it is utterly useless. Hence to prove the uniformity of nature we have to assume it. This is a clear-cut case of arguing in a circle—of assuming the very thing we are trying to prove. It is just as though we should try by means of argument to change the point

[3] From *Science, The False Messiah* by C. E. Ayres, copyright 1927 by The Bobbs-Merrill Company, 1954 by C. E. Ayres, reprinted by permission of the publisher. P. 129.

of view of someone who does not believe in the legitimacy of argument; we should hardly expect to succeed. Or it is as though we should argue that we cannot know how objects in the external world really appear because we must always see them through our own eyes, not realizing that in admitting this we are granting that we *do* know how *our own eyes* appear in the external world —which we cannot without the use of mirrors, cameras, or other instruments—and are thus contradicting ourselves. If the presuppositions of science are necessary in order to justify the method of science we cannot turn about and use the method of science to establish its presuppositions.

But if the scientific method cannot be used to establish the presuppositions of science is there any method which *can* be used? There seem to be only two remaining possibilities. We might say that the presuppositions of science are *self-evident*, or *obvious*, or *intuitively true*. (We may use these as equivalent expressions.) But we found in Chapter 7[4] that there are some difficulties with this notion. The postulates of Euclidean geometry cannot be intuitively true if the postulates of some other geometry contradict them, for these other postulates must also be intuitively true. The statement that a part is always less than the whole which contains it is not self-evident, for in the case of certain infinite wholes the part is equal to the whole. It is not self-evident that a surface always has two faces, for if one end of a strip of paper is rotated through 180° and the two ends are then pasted together the resulting surface has only one face, since one can start at any point on the surface and arrive at a point "on the other side" without "going over the edge." Evidently truths which *seem* to be self-evident may not really be so at all. Indeed, some of the presuppositions of science do not even *seem* to be self-evident. While it is apparent that nature is uniform in the sense that there are many regularities, there are also many irregularities and in this sense nature is not completely uniform. Again, while it is true that nature is objective in the sense that we do not usually change a thing *merely* by knowing it, nevertheless in some cases we *do* transform it by this process;[5] for example, one of the methods which a psychoanalyst employs in order to learn something about a patient is to talk with him, but in this process of talking with

---

[4] See above, p. 143.
[5] See above, p. 39.

the patient the doctor frequently does change him in significant ways. All of this shows that the presuppositions of science cannot be obviously or intuitively true.

The remaining possibility is that the presuppositions of science are "true by definition." If we adopt this approach we can say that nature *must* be uniform, knowable, and subject to causal laws simply because if some part of it proved to be not possessed of these characteristics we should deny that it is nature. (We should probably say that it is supernatural.) But this does not solve our problem. For it would mean that an apparent exception to a scientific law might be either a *genuine* and *natural* exception, in which case the law would have to be altered or rejected because nature is not what we had supposed it to be, or it might be a *supernatural* intervention, in which case the law would remain natural and would require no revision. But we should not be able to tell in any given case *which* decision to make. Or if nature were, by definition, completely knowable, what should we do about our failure to discover the cause of cancer? Should we say that cancer is not a natural happening since we have been unable to explain it, or should we persist in searching for its cause until we are able to fit it into the pattern of nature? Definition is an arbitrary act, and we cannot give the world the properties we want it to have merely by so defining it. In fact, to say that the presuppositions of science are true by definition is to invite the question: *Why* are they true in this way? To know that they are true by definition itself requires justification. No amount of word-juggling can help us out of this difficulty.

Where, then, do we stand? Belief in the presuppositions of science cannot be justified by means of the scientific method, nor by claiming that they are self-evident, nor by any kind of word-manipulation. There remains only one possibility: they are true because they work. And when we say this we claim that if and only if they are assumed to be true can science be a more satisfying experience in the sense that it leads progressively to knowledge which is more and more certain, more and more clear, more and more extended, and perhaps even more and more useful. This is frankly to admit that the belief in the attainability of truth, the objectivity of nature, the rationality of the world, and the uniformity of natural processes is warranted simply because without it the intellectual enterprise would not continue; or would continue only as a less satisfying experience than it could be. Man is so constituted that he must pursue truth, and probably will continue to do so even though he is not able to justify his behavior. But he will be much happier in this activity if he can know the framework within which it lies and the limitations to which it is subject. Only if he has some conception of what can and what cannot be known can he steer a middle course between the Scylla of self-depreciating humility and the Charybdis of arrogant dogmatism. In no other way can he understandingly exhibit the scientific spirit. This solution to the problem of the justification of science is commonly called the "pragmatic" solution,[6] and will be designated in this way in the following pages. It does not commit one to the acceptance of pragmatism in any other of the many meanings of this term.

4. The Presuppositions of Religion

Now let us turn, after this long excursion, to religion. The presuppositions of religion are those beliefs which, if true, justify religion. They constitute its creed. But they are not the conditions of religion in the sense that they are consciously present before religion occurs; on the contrary they are called forth by the fact that, sooner or later, we begin to have misgivings about our religion. We find it less satisfying than we had hoped, and we begin to search for something which will restore this quality. The creed of religion involves, as does the creed of science, two elements. First, piety is one of the supreme values of life. This means that it is worth striving for even though such activity may require sacrificing other values. Second, piety is attainable at least by approximation. "Religion" is commonly defined in such a way as to make these two presuppositions equivalent to the assertion of the existence of "God," in some of the many meanings of this word. Since our purpose is not to make a thorough analysis of

[6] Others who have accepted this solution are the following: H. Feigl, "De Principiis non Disputandum . . . ," *Philosophical Analysis*, ed. Max Black (Ithaca: Cornell University Press, 1950), pp. 119–156; R. B. Braithwaite, *Scientific Explanation* (Cambridge: Cambridge University Press, 1953), Chap. VIII; Hans Reichenbach, *Experience and Prediction* (Chicago: University of Chicago Press, 1938), Chap. V; and Ernest Nagel, "Probability and Degree of Confirmation," *Philosophy of Science*, ed. Arthur Danto and Sidney Morgenbesser (New York: Meridian Books, Inc., 1960), pp. 253–265.

religion we may accept this identification. We may then say that belief in the existence of God constitutes justification for the pursuit of piety in essentially the same way that belief in the objectivity and rationality of nature constitutes justification for the pursuit of scientific truth.

Writing in the *American Scientist*,[7] Harold G. Cassidy points out the striking similarity between the creed of the scientist and the creed of the religious man. Suppose, he suggests, we were to modify Archibald MacLeish's play, *J.B.*, written on the theme of the Book of Job, by transforming Job from a businessman into a scientist. We might then assign to him Galileo's task of discovering the law governing the rate of speed of a ball rolling down an inclined plane. Let us further suppose that every time Job performs the experiment he not only detects different relations between the distance and the time but discovers that the ball sometimes stops midway in its course and even on occasions rolls up the plane instead of down. Of course this will irritate Job since he firmly believes in the uniformity of nature. But he does not give up, and in spite of the trials he is undergoing he persists until he finds a formula which explains the erratic behavior of the ball. Just as the biblical Job retained his faith in the goodness of God even when he was experiencing persistent and continued torture, so the scientific Job retains his faith in the legality of nature even when its behavior seems to be completely haphazard. The continuing faith of medical research in the existence of a cause for cancer, in spite of repeated failures to discover it, is of this same kind.

One of the charges very commonly brought against religion, especially in an age of science, is that it rests on presuppositions which it cannot justify. Every religion, so the attack runs, if it is anything more than a spontaneous and childlike outpouring of the soul, demands a creed—a group of beliefs about God, man, and the world. This creed, supposing religion to be more than a passing fancy, must have some sort of intellectual guaranty. But no such assurance can be found. Religion rests only on such insecure foundations as divine revelation, authority, and emotional need. Science, on the other hand, can stand on its own feet; it

[7] "The Problem of the Sciences and the Humanities—A Diagnosis and a Prescription," *American Scientist*, Vol. 48, No. 3, 383–398.

requires no guaranty. Religion must therefore disappear in a scientifically oriented society.

The first step in reply to this charge is the frank admission that religion does, for many people, demand justification.

Carlyle has argued agains this claim. He pointed out with some effectiveness that religion should be "let alone," and that we should resist the tendency of our rational faculties to enter this domain. Except we become as little children we cannot enter into the Kingdom of Heaven. Instead, therefore, of fostering the critical attitude in religion we should discourage it. Even in the sphere of inquiry we should remember the discontented Socrates. Only the value experiences at the primitive level are finally satisfying; as soon as analysis and criticism arise the experiences melt away.

The element of truth in this claim covers the error which it also contains. Unquestionably for many people an unreflective and uncritical value experience is quite satisfying. Naïveté is, from many points of view, a desirable attitude. But we should remember that it is a value only for those who are sophisticated, not for the naive. The individuals who prefer, say, an unexamined religion to an examined one are not really naive, since, in recognizing the possibility of an alternative to an uncritical religion, they have become critical about it. Even though they may claim that an unreflective religion is "good enough for them," the well of satisfaction has been polluted and their simple religion can no longer please them in the same way or to the same degree. The only person who can be really satisfied with an uncritical religion is one who is so naive that he has never heard of a critical religion – one for whom the choice between a critical and an uncritical religion would be strictly meaningless.

Furthermore intellectual growth in value appreciation does not necessarily destroy the enjoyments of the prerational stages. The essential advantage of the reflective approach lies in the fact that the individual with the more critical and discerning value enjoyment has, in a sense, all that he had in the days when he was intellectually less mature or, at least, can frequently with a certain amount of effort return to that pristine stage. But he has, in addition, the increased appreciation which is the natural outcome of a wider range of value experiences. However intense the lower level satisfactions may be, they are usually adequate only to the simpler situations of life and tend to break down in the face of the complicated problems which confront the individual in adulthood. When conflict arises choice becomes necessary, and the most satisfying choice is one which is reflective. Thus growth in value appreciation is identical with increase in the ability to make a wise choice. Such choice is dependent upon knowledge of the nature and presuppositions of the discipline and field in which the choice is made. In general the greater the range of knowledge, the more adequate the choice will be, and the more satisfying the experiences. The problem of the enhancement of enjoyment in the area of value experience is, therefore, the problem of the progressive disclosure and justification of its presuppositions.

We find ourselves, consequently, led to the conclusion that religion does sooner or later demand its own justification, and rightly so. While we should never ask for the justification of religion if we were not first religious, we are less likely, in view of our continually broadening experience, to remain satisfied with being "just religious." We feel strongly the urge to integrate our religion with the cognitive aspects of our lives and establish on intellectual grounds a creed, a theology, or a metaphysical theory on the basis of which the continuation of the religious activity may be rendered justifiable.[8]

If we grant that religion is entitled to ask for justification is it true, as the charge claims, that no such justification can be found? Must religion be abandoned as an "unscientific" attempt to meet the problem of happy living?

There are two possibilities open to religion. First, it might justify its creed by science. God would then be a hypothetical being whose existence is to be established by the usual techniques of science — observational, experimental, inductive, and deductive. This would not involve arguing in a circle as it did in the case of the justification of science, for in that case science was being justified by science while here religion is being justified by science. Now science may or may not be able to prove the existence of God. Many scientists have argued that it can; many others have argued that it cannot; some have even argued that it can prove that God cannot exist. The disturbing thing about the scientific proofs for the existence of God, taken as a whole, is that God turns out in each case to be a different sort of being. For Sir James Jeans God must be a mathematician; for Sir Arthur Eddington he cannot be a mathematician; for those who urge the "design" argument God must be a great engineer, like a skillful watch-maker; and for a limited group who in recent years have been impressed by the need for expressing more and more of our scientific laws statistically, God determines the destiny of the universe by playing dice.

But even if science *can* demonstrate the existence of God we should be quite clear as to what this implies. We are entitled to

[8] A. Cornelius Benjamin, "The Justification of Religion," *Journal of Religion* (The University of Chicago Press), XXVII, No. 1 (January, 1947), 30–31. Copyright 1947 by the University of Chicago.

believe in the existence of God for the same general reason that we are entitled to believe in the existence of electrons, namely, that they are both truths established by science and justified by the presuppositions of the scientific method. But can *these* presuppositions be established? If they cannot be justified within science except by arguing in a circle; if they are not self-evident; and if they are not true by definition, then we must fall back on the pragmatic guaranty: they are true because they work. They are acceptable because, assuming them, we obtain more accurate knowledge, more certain knowledge, and more comprehensive knowledge. These are important cognitive values, and if one feels that his religion is strengthened by this grounding he is entitled to do so. But what he is doing is reducing religion to religious faith, religious faith to faith in science, scientific faith to faith in the presuppositions of science—whose only guaranty is their workability.

On the other hand, religion may justify its creed directly, without employing science as an intermediary. This may be done by showing that belief in the existence of God, or belief in some conception of the nature of the universe and man which will provide the same sort of drive that belief in the existence of God does—*whether or not this belief can be scientifically grounded*—has the same relation to the religious experience that belief in the uniformity of nature has to science itself as a mode of inquiry. The presuppositions of both science and religion are guaranteed by the fact that they increase our understanding of these experiences, they help to carry us through periods of despair and disillusionment, and they enhance both the pleasure of the pursuit and the joy of the attainment. Just as a science whose presuppositions have been disclosed and clarified, even though they cannot be proved, is a more satisfying experience than a blind science, so a religion whose creed has been formulated, even though it cannot be established, is more rewarding than an innocent religion. The scientist certainly gains confidence and comfort in the knowledge that his presuppositions are at least plausible truths and that any natural events which appear to violate them can usually be taken care of without abandoning them. In much the same way the religious man, even though he is unable to prove that man and nature are modifiable in the direction of perfection, does derive satisfaction from knowing that such a conception of human nature and the universe is at least believable and that it is sufficiently flexible to take care of those data in his experience which seem to run contrary to it. The presence of evil in the universe does not lead the religious man to abandon hope in the perfectibility of man any more than the failure of the pathologist to find the cause of cancer persuades him that there are uncaused events in nature.

It has not been the purpose of this chapter to "save" religion from attacks by science. Indeed, if science and religion are both value pursuits, resting on unprovable assumptions, how could either be saved from attack by the other? They are on the same level, and apart from an evaluation of values, whose consideration will occupy us in the last chapter, neither takes precedence over the other. The fact is that religion occupies that same status as the pursuit of aesthetic values, moral values, political values, economic values, and values of health and play. Religion was selected from among these for special consideration in this chapter simply because in the history of thought it has so frequently been asked to state and justify its presuppositions. The other value activities can also properly be asked to justify themselves, and their attempts to do so would probably follow the general lines laid down in the case of religion. Furthermore, the chapter has attempted to show that science, which has also rarely been asked about its presuppositions—what they are or whether they can be guaranteed—can be properly asked the same question. Thus far it has provided no satisfactory answer. The pragmatic solution, proposed in this chapter, is only one of many ways in which the problem may be handled. Eddington has expressed both the problem and its solution very clearly: "In the age of reason, faith yet remains supreme; for reason is one of the articles of faith."[9]

[9] A. S. Eddington, *The Philosophy of Physical Science* (New York: The Macmillan Company, 1939), p. 222. Some suggested readings on the topic of this chapter are the following: Albert Einstein, "Science, Philosophy and Religion," *Readings in Philosophy of Science*, ed. P. P. Wiener (New York: Charles Scribner's Sons, 1953), pp. 601–607; J. A. V. Butler, *Science and Human Life* (New York: Basic Books, Inc., 1959), Chap. 14; L. A. Reid, "Religion, Science and Other Modes of Knowledge," *The New Scientist: Essays on the Methods and Values of Modern Science*, ed. P. C. Obler and H. A. Estrin (Garden City: Anchor Books; Doubleday & Company, Inc., 1962), pp. 239–265; J. Arthur Thomson, *Science and Religion* (New York: Charles Scribner's Sons, 1929); and R. A. Millikan, *Evolution in Science and Religion* (New Haven: Yale University Press, 1930).

Education Through Science

The widespread interest in higher education today is the result of many factors. On the practical side, snowballing enrollments will increase the demand for buildings, staff, equipment, and improved teaching instruments. The development of automation will increase our leisure. The demand for highly trained scientists may well unbalance our curriculum. The advent of Sputnik has made us afraid in our own world. Extremist ideologies at home and abroad have rent us asunder. Small wonder that there are many people today who are asking how we can best prepare our children for the kind of world in which they can be expected to live happily. At least thirty years ago educators were worrying about counteracting the effects of a curriculum devoted largely to such courses as homemaking and personal adornment, which were obviously "low in intellectual content," or to courses in science, which were clearly too specialized to prepare the student for a life in which his major choices would lie in the areas of art, religion, morality, politics, and social behavior. Throughout the period the key phrase has been "general education," and controversy has centered mainly around the place of science in such training. Educators seem to agree that science plays an important role in general education, but *why* it does, *how much* science, *what* science, where science should *occur* in the curriculum, and how science should be *taught,* are problems producing great volumes of dispute but only a trickle of ideas.

1. Science and Culture

One of the most serious obstacles to the solution of such problems is the deep-seated conviction that science and the humanities are opposed to one another. The English scientist and novelist, C. P. Snow,[1] whose recent book, *The Two Cultures and the Scientific Revolution,* has already been mentioned in the Introduction, deplored the fact that the humanists (who are honored by being characterized as "cultured") and scientists (who are consequently described as "uncultured") can scarcely speak to one another, since they have neither a common language nor a common area of interest. The specialization of science into minute areas, and its practicalization into gadgets instead of ideas has raised even higher the barrier which separates them from the men of arts and letters. Indeed, it is a real disaster when questions as to use of nuclear weapons for waging war must be decided; for the scientists who know the potentialities of the weapons do not know the tactics of political diplomacy, and the politicians and military men who must make the final decision are unaware of the powers and dangers of the weapons.[2] Effective action obviously requires that the two groups be brought together. Snow's suggestion that we henceforth speak of the "two cultures" rather than the "culture of the humanist" and the "lack of culture of the scientist" is certainly praiseworthy and pertinent. But it does not meet the practical problem of cross-fertilization of the two groups.

In the field of education itself, there has often been a strong feeling that science and the humanities are mutually exclusive.

---

[1] New York: Cambridge University Press, 1961.

[2] " 'No modern statesman,' says Eban, for ten years Israel's Ambassador to the United States and her chief delegate to the United Nations, 'can afford to be scientifically illiterate.' " Warren Weaver, "A Great Age for Science," *Goals for Americans: The Report of the President's Commission on National Goals* (Englewood Cliffs, New Jersey: Prentice-Hall, Inc., 1960), p. 103.

Both, it is readily admitted, have their contributions to make to the development of the well-rounded personality, and a curriculum which includes the sciences but neglects the humanities, or one which includes the humanities but neglects the sciences produces a narrowness of outlook which is foreign to the aims and ideals of a broad education. But if we forget, momentarily, the liberalizing effects of education and ask what it can provide in the way of assuring the most effective use of science in this technological age, the problem becomes much more vital. Progress in science affects human values and changes in human values affect science. This interaction requires, on the one hand, that we produce more scientists and better educated scientists. And because competence in science can no longer be acquired by taking a few courses in the field, more *time* must be devoted to science in the curriculum. On the other hand, since many of our most vital decisions in the immediate future will be policy decisions—decisions concerning international trade, aid to underdeveloped nations, the cold war, and colonialism—we must train men more effectively in foreign languages and in the areas of social, political, economic, and moral values.[3] Granting a finite curriculum, who is to give way to whom? Do we want men well trained in the sciences who are incapable of making policy decisions, or do we want men informed in values and skilled in diplomacy who are ignorant of the many instruments which science has provided for helping us achieve our ends?

The obvious answer is that we want men well trained in both areas—if we can get them. How to bring this about is a critical problem for educators. Presumably what they will have to do is either to extend the period of education or to squeeze some of the water out of courses in the elementary and secondary schools. But this is not the problem I wish to consider. I am concerned with attempting to provide some sort of principle on the basis of which we may "reconcile" the sciences and the humanities by determining the respective roles which they play in the educational scheme.

[3] A scientific friend recently tried to convince me that while it was absurd even to think of the humanist acquiring a "working" knowledge of science, the reverse was not true. For example, he suggested, if a scientist were confronted with an ethical decision—say, whether or not to release some classified information without authorization—he need only set aside a weekend, look up the article on "Ethics" in a good encyclopedia, and return to his office on Monday with his problem solved. I had to remind him that Aristotle did not write his *Ethics* on a weekend.

If we are to include the sciences and the humanities in the curriculum we ought to know why they should be required, what each contributes each in its own way, in what sense (if any) they are opposed to one another, and how each of them affects the other. We cannot have an integrated curriculum unless we have an integrating goal. We cannot presume to give the students a *general* education unless we tell them what (in general) the humanities are, and what (in general) the sciences are.

The administrative and curricular problems involved in introducing subject matter of this kind into our institutions of higher education are complicated, and cannot be considered here. Many colleges and universities attempted to meet the problem twenty or thirty years ago by introducing so-called "general" or "integrating" courses into their programs. These usually consisted of an amalgamation of several courses—say, physics, mathematics, chemistry, and geology—into a sweeping course in the physical sciences, or a combination of botany, zoology, and physiology, into a course in the biological sciences. But frequently they did not accomplish the end for which they were designed, since they merely substituted a few facts from many fields for many facts drawn from a single field, and there was usually little attempt to achieve a genuine integration of the facts, to uncover common methods, or to examine the historical or social environment of science. What was really needed was radically new courses, or greatly modified old courses, in the history, psychology, sociology, and philosophy of science.

2. The Cultural Fringe of Science

If science, properly considered, is to play an effective role in higher education, opportunity must be offered to look at it from a new vantage point. Science as a human enterprise is not a mere collection of facts and formulas, nor is it the manipulation of gadgets and the solving of equations. It is something with a philosophy, a history, a sociology, a psychology, even an economics and a geography. It represents a phase of the developing human spirit, both cultural and individual, and must be understood in terms of the social factors which produced it and the social transformations which it itself is producing, and in terms of the personalities who have both created it and been created by it.

Naming is always a dangerous process, since it encourages us to attach to an object all of the attributes which belong properly only to the name. But in spite of these risks I venture to propose that what we have been talking about be called the "cultural fringe of science." The periphery of science is not strictly science, yet the two are certainly closely related, and an interest in science can be channeled, broadened, or perhaps transformed into an interest in the periphery of science. That this border of science is cultural or humanistic has already been shown, and will be further developed in what follows. One might be tempted to say that what we want is a device for luring the scientifically minded students into the humanities, but this is not at all the case. A scientist who yielded to such a temptation would probably not have been a good scientist, and would end by becoming a bad humanist. What is desired is rather a method for capitalizing on, developing, and satisfying the cultural interest which most scientists possess, and which they must have to an increasing degree if society is to survive. Only by focusing this interest on those problems which take their origin in science itself but are not met by the narrow and specialized consideration of scientific subject matter, can this be accomplished. We do not want scientists who know *less science* and *more about science*; we desire scientists who know both science and about science.

3. The Philosophy of Science

First on the cultural fringe of science we find the *philosophy of science*. If our view developed in Chapter 9 is correct, science, as a mode of inquiry, is not opposed to the humanities but is one of them. Since this point has been developed at length all that we need to do is to re-emphasize scientific truth as one of the goals pursued by man in essentially the same way that God, beauty, and the moral good are, and—under certain circumstances at least—for essentially the same reasons. Science is a human enterprise, a valuational response, an activity which in the truest sense of the word may be said to be concerned with spiritual values and with the "highest" in man. When we think of the sciences we tend to forget the behavioral sciences. The scientist seems to be concerned with mere matter and hence to be neglectful of consciousness and purposiveness as aspects of the world. But while it is true that the natural scientist studies matter, the studying is not matter; it is an activity of the spirit filled with significance and value. And when we include behavioral sciences as well, the preferential aspect of science becomes more pronounced. To refer again to C. P. Snow,[4] he talks constantly about the "gulf of mutual incomprehension" which separates the literary intellectuals from the scientists. He admits that the number "2" is a very dangerous number and that anyone who divides things in this way ought to be regarded with suspicion; in fact he "thought a long time about going in for further refinements" and finally decided against it.[5] But here, right at his hand, are the behavioral scientists, who attempt to bridge the gap between the intellectuals and the scientists, yet whom he never calls upon in the entire book to help him out of his plight. Scientists, he seems to forget, *behave*, and it is the philosopher (an intellectualist!) who describes and tries to give logical structure to that behavior.

That science has close and intimate relations to the philosophy of science seems undeniable. In fact most scientists recognize this and more of them would be willing to admit it—if the study were not called "philosophy." Common sense demands recognition of the intense emotional reactions which can be aroused by mere names. Philosophy has often been in disrepute. It has used so freely (though perhaps unavoidably) the very techniques and methods which have been rejected by science, and it has stood so long for uncontrolled speculation, fuzzy thinking, and futile argument, that it has become synonymous in the minds of many scientists with pure quackery. This aversion to the subject, based largely on an antagonism to a mere word, is much less intense than it was, say, twenty-five years ago. Most colleges and universities today have established courses in the philosophy of science, the literature in the field has increased rapidly, and the American Association for the Advancement of Science now has a section devoted to the history and philosophy of science. Perhaps if the subject were called "analysis of the sciences," "criticism of the sciences," or even "meta-science" there would be less hostility toward this new discipline, which, as a matter of fact, has never

[4] *The Two Cultures and the Scientific Revolution* (New York: Cambridge University Press, 1961), p. 4.

[5] *Ibid.*, p. 10.

claimed to be a substitute for science and never tried in any way to "undermine" it.

Conant is outspoken in his criticism of the philosophical approach to the study of the scientific method. In spite, he says, of the lack of competition from other fields, "I doubt if the philosophical treatments of science and scientific method have been very successful when viewed as an educational enterprise. . . . I am inclined to think that, on the whole, the popularization of the philosophical analysis of science and its methods has led not to a greater understanding but to a great deal of misunderstanding about science."[6] The lack of justifiable grounds for Conant's prejudice is clearly indicated in the example which he gives in support of his belief—Karl Pearson's *Grammar of Science*. This was indeed a popularization of the scientific method, read very widely, and debated extensively in "discussion clubs" throughout the country. But Conant fails to mention some very important considerations. Pearson's book was published in 1892, a year when the philosophy of science was still in its infancy. Furthermore, while *Grammar* is commonly considered to be philosophical in character (Pearson wrote other books in the same vein), the author was not primarily a philosopher but a scientist, whose *Science of Mechanics* is a standard work in the field, and who himself made important contributions to the science of biometry; he was therefore not ignorant of science. Finally, and *most* important, Pearson was an ardent positivist and firmly believed that the task of science was essentially classificatory and descriptive, rather than explanatory or deductive. Thus he belonged to a *school* of philosophers of science, with whom Conant happens to disagree. I have attempted in the previous pages to show just what there is to be said for and against the descriptive view of science, and my hope is that these words, in spite of having been written by a philosopher and in spite of Conant's pessimism, have offered some enlightenment to the reader.

No doubt the antagonism often shown by the scientist toward the philosopher when questions of scientific method come up for debate is caused by the philosopher himself. Certainly no scientist should object to having his attention called to the fact that science rests upon assumptions. But the approach of the philoso-

[6] J. B. Conant, *On Understanding Science* (New Haven: Yale University Press, 1947), p. 28. Reprinted by permission of the publisher.

pher in this task of uncovering presuppositions is sometimes made in an indiscreet manner. It frequently takes the form of telling the scientist that his entire structure is built only on sand and that unless he promptly starts some shoring and strengthening operations it will sink into the sea. To this the scientist can justly reply that science has been doing very well for several centuries, thank you, in spite of the general neglect of these considerations, and will probably continue to do so. He can even admit that he need concern himself with such matters only when, as in the case of organic evolution, non-Euclidean geometry, relativity physics, and nuclear fission, he is compelled to do so by the progressive development of science itself. No scientist likes to be told, especially if it is the philosopher who tells him, that he has not properly thought through his procedures. A much more fruitful approach, it seems, would be to place all such considerations outside science itself and locate them on its cultural fringe. Then the scientist who does not have time to examine his presuppositions but who fears that the very survival of science depends upon such activities, can salve his conscience by turning the job over to the philosopher. And the scientist who does have time and is both genuinely scientific and incurably philosophical, is provided with some justification for his week-end speculations. There is no reason why one should not examine his presuppositions in response to the same urge which induces him occasionally to read a good book, or attend a symphony concert. The point is that in the area which immediately surrounds science these paths which lead out of science into broader fields are present in great numbers. It seems a pity, therefore, that the educational scheme has not recognized them.

One of the strongest reasons for emphasizing the teaching of method is the ease with which students in science may be led to develop an interest in its procedures and techniques.[7] Beginners in science can usually be presumed to know two things about science: First, science has produced results in the history of

[7] An interesting report in *Science Education News* (American Association for the Advancement of Science), November, 1963, states that the newly established Commission on Science Education has met with great success in teaching physics to students in the elementary schools, by giving them an inclined plane, balls of different weight and of different material, some solid and some hollow, and encouraging them to "experiment" with a minimum of guidance in order to show by actual practice how Newton's laws of falling bodies can be derived.

thought and is therefore an example of a thoroughly practical enterprise. Second, its method is substantially that of common sense and is consequently easy to understand, at least in its outlines. But this appeal is completely lost when the teaching is bad.

> Such is the preoccupation in the teaching of science with specific subject-matter and with specific technologies, that many students, although they major in some particular science, have no conception of the scientific method as a generally valid approach to the problems of this world. Science is generally thought of as a type of *subject-matter* rather than as a *method* of study. . . . In short, if we wish to remedy the present failure of our schools to acquaint students with even the elements of scientific method, we must establish and require, from the grades through high school and college, courses definitely calculated to acquaint everyone with the broader meanings and methods of science.[8]

To be sure, one of the first jobs of the teacher should be to show students that science has in fact blundered often in its history, that it is now and probably always will be confronted with problems which it is not able to solve, and that its method is often highly complicated and involved. Unquestionably many students, as a result of an initiation into the scientific method, will, like Plato's youths,[9] try it out on everybody and everything with the expected ridiculous results. But if they do this they have not been taught effectively, for a proper understanding of the scientific method involves a conscious recognition of its limitations. The point is, simply, that in science we have a clear-cut example of the mind at work, and the student may be induced, if his interest can be sufficiently aroused, to pay some attention to this fact. In doing so he is opening for himself a wide variety of exploratory avenues all of which lead, in the strict sense of the word, away from science and into the broader fields of logic and theory of knowledge. Thus the way is clear, if the spirit is willing, for the development of that wider outlook which is characteristic of the cultivated and cultured man.

But two other aspects of the philosophy of science encourage students to go into the broader cultural areas. Both of these were mentioned in Chapter 9. On the one hand, there is philosophy as the most basic of the sciences—that subject dealing with the

---

[8] G. A. Lundberg, *Can Science Save Us?* (New York: Longmans, Green & Company, 1947), pp. 66–67. Courtesy of David McKay Company, Inc.

[9] *Republic* (Jowett edition; New York: Charles Scribner's Sons, 1887), sec. 539.

most general of the categories underlying all things. The pyramidal arrangement of the sciences given at this point destroys the isolation of the individual sciences and almost compels us to see them as a unified structure. It is this pattern on which all "reductionist" and all "emergence" theories are based. Reductionism[10] involves a "nothing but" view of the hierarchy of the sciences; consciousness is nothing but life, life is nothing but matter, matter is nothing but abstract forms, and forms are nothing but the categories. Emergence theories always involve a "something more" relationship; abstract forms are something more than the bare categories; matter is something more than abstract form; life is something more than matter; and consciousness is something more than life. All of the controversies concerning mechanism and vitalism, and between behaviorism and mentalism lie inherent in this structure.

On the other hand, without the cultural fringe each science tends to be unrelated to all the other sciences. The high specialization in the sciences tends to support this outlook, and, indeed, further separation is to be expected in the future in view of the increasing range of material to be covered by the sciences. But what is often forgotten is that to know that A is independent of B requires us to know something about *both* A and B. Laplace's famous remark that as an astronomer he did not need the hypothesis of God illustrates both the legitimacy and the illegitimacy of isolating the sciences. One who was merely an astronomer could not make such a statement; to know that he did not need God, Laplace had to know who God is. Whether, therefore, the sciences be considered independent of one another, as in the presumed case of astronomy and theology, or dependent on one another, as in the well-recognized case of physics and mathematics, each science overlaps with every other. Furthermore, recent trends show an increasing emphasis on cooperative efforts among the sciences, particularly in areas of applied science. In the use of motivational research manufacturers of certain products are attempting to increase their sales by seeking advice from advertisers, television programmers, psychoanalysts, sociologists, and econo-

[10] Ernest Nagel, *The Structure of Science* (New York: Harcourt, Brace & World, Inc., 1961), Chap. 11; and Richard Jessor, "The Problem of Reductionism in Psychology," *Theories in Contemporary Psychology*, ed. Melvin Marx (New York: The Macmillan Company, 1963), pp. 245–255.

mists. In the use of operations research, which began in World War II as a strategic and tactical method for solving problems of winning the war by "joint" conferences of men with highly diverse training and interests, we have another example of the way in which complicated problems may be solved by a pooling of information drawn from widely separated sources. This method is now commonly used in improving the efficiency of industrial concerns, fund-raising drives, political campaigns, and other enterprises where several goals are involved and the intermixture of means and ends is a highly complex matter. Any consideration of the interrelations of the sciences more or less forces one to step outside his own field, and the study of science becomes something more than science. In this way a strictly technical and specialized interest becomes broadened into a philosophical and humanistic interest without sacrifice of the narrower interest and without prejudice to the broader.

4. The Psychology of Science

The *psychology of science,* as we saw in Chapter 8, where we discussed only a small part of the problem, is a second neglected area. When we study science today we do not ordinarily study scientists. This is fitting and proper; the scientist as a man does not enter into science in the strict sense of the term. Science as taught today is the distillation of hundreds of minds. Errors and idiosyncrasies have disappeared; there remains only the Abstract Scientist as a reasoning machine. But though this is a convenient fiction it is hardly true to the facts. The history of science is the history of scientists—of men like you and me, who are members of a particular historical and cultural group, and they accomplish their results because of, or perhaps in spite of, this fact. The history of science, for example, cannot be written apart from the history of men of genius, for we know little or nothing of that flash of insight[11] by which great ideas are called into being. To be sure, the psychology of science should not be *identified* with the science of which it is a study. Science need not know how a man received an inspiration; it need only ascertain what the idea was and whether it turned out to be correct. Consequently the

[11] See above, Chap. 5, especially sec. 5.

investigation of such matters is not ordinarily given in courses in science, but it does belong in the periphery of science because it tells us what kinds of people scientists are and how science, as a matter of fact, achieves its end.

5. The Sociology of Science

The *sociology of science*[12] certainly includes much of the history and biography of science, and therefore may not exist as a separate field of study. But it is not science itself—at least in the sense that it is not to be identified with the science which it studies. Looked at from this point of view science is a phase of human culture, having relations to all other social phenomena—to religion, art, government, morality, education, health, and recreation. Furthermore, as a social phenomenon science is an organization or an institution having members, bylaws, prerogatives, and vested interests. For example, scientists are finding that their code of ethics of a hundred years ago, which required them to publish immediately the results of research, has now been largely abandoned. Results which have been obtained through industrially sponsored research are the private possession of the company which paid for them; results pertaining to missile research which has been sponsored by the national government are immediately classified. This is a new kind of science, not merely in the sense that it is practical rather than pure, but also in the sense that it profoundly affects the motivational factors which induce scientists to pursue a scientific career.

6. The History of Science

The last of these fields on the cultural fringe of science is the *history of science*. Too often it is forgotten that science, like everything else, has a history. No human enterprise can be thoroughly understood without some reference to the temporal period in which it is found, and the antecedents from which it emerged. This does not mean either that the history of science is science, or that the best way to teach science is through its history. Auguste Comte was right in cautioning against these identifications.

[12] This topic will be discussed in greater detail in Chapter 13.

But it does mean that a proper understanding of science, especially in its tendency toward dogmatism and absolutism, can be evaluated through a study of its history.

> There are two ways of probing into complex human activities and their products: one is to retrace the steps by which certain end results have been produced, the other is to dissect the result with the hope of revealing its structural pattern and exposing the logical relations of the component parts, and, incidentally, exposing also the inconsistencies and the flaws. Philosophical and mathematical minds prefer the logical approach, but it is my belief that for nine people out of ten the historical method will yield more real understanding of a complex matter.[13]

With this evaluation of the superiority of the historical over the "logical" method for teaching scientific methodology, I should strongly disagree. True, it is only recently that we have tried the historical approach; most histories of science[14] are not written with the aim of uncovering methods and patterns, and there is consequently very little textual material which is available to the student for this type of approach. Several American universities now have joint departments in the history and philosophy of science. Here the retarding factor in preventing more rapid expansion is the poverty of men who are properly trained in the two fields. But both the history of science and the philosophy of science stand to profit greatly by this new association. Students from the humanities, who are generally not well informed in the sciences, are attracted by the double fact that sci-

---

[13] Conant, p. 27. Reprinted by permission of the publisher.

[14] W. S. Jevons, *Principles of Science* (London: Macmillan & Co., Ltd., 1907), though not a history of science, abounds in illustrations drawn from early science. J. Bronowski, *The Common Sense of Science* (Cambridge: Harvard University Press, 1958) is an excellent presentation of the scientific method as seen in its development. A quotation from this book is relevant: A knowledge of the history of science "gives us the backbone in the growth of science, so that the morning headline suddenly takes its place in the development of the world. It throws a bridge into science from whatever human interest we happen to stand on. And it does so because it asserts the unity not merely of history but of knowledge. The layman's key to science is its unity with the arts. He will understand science as a culture when he tries to trace it in his own culture." P. 3. Reprinted by permission of the publisher and Heinemann Educational Books Ltd., London. "The Structure of Scientific Revolutions," by Thomas S. Kuhn, *International Encyclopedia of Unified Science*, II, No. 2, is a valuable recent addition to the literature. Other references are: J. B. Conant, *Science and Common Sense* (New Haven: Yale University Press, 1951); *Theories of the Scientific Method: Renaissance to the Nineteenth Century*, ed. E. H. Madden (Seattle: University of Washington Press, 1960); and Cecil J. Scheer, *The Search for Order* (New York: Harper and Brothers, 1960).

ence speaks with authority and that the scientific method, at least in its essentials, can be readily understood by the layman. On the other hand, as Kuhn[15] points out, students from the sciences, while they are the most rewarding group to teach, often prove frustrating because they already "know the right answers" and are therefore unable to understand early science in its own terms. But both types of student, when they learn to their amazement that early scientists believed quite as firmly in the caloric theory of heat as present-day scientists do in the molecular theory, soon recover from their shock and emerge with a new appreciation of what it means to confirm a scientific theory. They learn that social, economic, and political circumstances condition the growth of science, and that science, in turn, determines social, economic, and political events. Thus from a retrospective knowledge of science they gain perspective.

The conclusions which we can draw from the examination of these areas of the fringe of science should be extended to the whole. Should science examine its own periphery? We have seen that this means: Should science become more than science? When the question is put in this manner the answer is negative: no science *must* ever examine its periphery in order to survive. However, suppose as educators we formulate the question: Should a scientist be more than a scientist? Here the answer is no longer a simple matter. It is certainly not an unequivocal negative. But if the principles of general education are correct the answer should be in the affirmative, for the question becomes: Should the specialist become more than a specialist? The thesis of this chapter has been that in the problems of the history, psychology, sociology, and philosophy of science we have an area of study where opportunity may be offered for the specialist to become more than a specialist.

To argue for an increased emphasis on the cultural fringe of science is not, of course, to disparage the emphasis on science. Quite obviously our very survival depends on producing as rapidly as possible a large number of highly competent scientists. Many institutions, in addition to the colleges and universities, are working in this direction. Television programs, both those that popularize science and those that offer college-credit courses in science,

---

[15] Kuhn, p. 166.

are invaluable aids toward this goal. The activities of the National Science Foundation and the provisions of the National Defense Education Act in making funds available for scientifically minded students who would otherwise find it impossible to continue their graduate studies are hopeful indications that we are making progress. Even the scientific toys for our youths, and the do-it-yourself kits for those of us who are not so youthful, encourage an interest in science and thus pave the way for better salaries for teachers of science, better laboratories for workers in science, and greater respect for those who only a few years ago were characterized as eggheads and longhairs. Nothing which has been said in this chapter is designed either to stop this movement or to slow it down. It has attempted rather to show that science is a cultural phenomenon, and that no cultured man, be he scientist or humanist, can afford to be ignorant of this fact.[16]

---

[16] Some of the ideas developed in this chapter were previously published by the author in an article entitled, "The Cultural Fringe of Science," *Journal of Higher Education,* XIV, No. 9, 455–462.

13

The Social Role of Science

The invention and use of the atom bomb in 1945 may go down in history as one of the most significant events of all time. This would not be because we were then fighting the most significant war of all time and the use of the bomb was the only method for ending it, for authorities are not agreed that this was the case; many argued at the time, and still argue, that Japan was on the verge of complete collapse and could not have lasted more than a few months. What the event told us was much more important than its probable effect on the war. It made abundantly clear, perhaps for the first time, that science is a social force of incalculable power. In a way we already knew this—at least in the sense that we knew science to be an important instrument of the good life. For we certainly had learned that science had given us greater physical comfort, better health, increased leisure, and more opportunities for cultural enjoyment. To be sure, we had also learned that science produced automation and automation produced unemployment, science invented automobiles and automobiles increased the hazards of travel, and science created farm surpluses and farm surpluses impoverished farmers. Thus we knew that science is also an important instrument of the bad life. But not until the atom bomb did we come to full realization that science could annihilate the human race. This was the grim truth which was brought to light on that eventful day, and which has been corroborated and strengthened by everything that has occurred since.

1. Science as a Social Institution

As a consequence of this discovery we are now looking at science in a different way: we are seeing it as a social institution.[1] To describe science in this manner is to say that it is a group of individuals having their aims and ideals, their ethics and their practices, their vested interests and pressure groups, and their own cults and factions. In this respect science is exactly like education, religion, politics, labor, and business. The goal of education is to increase the number of informed people and to improve the quality of the training which they receive; the goal of religion is to make more people more religious; similarly the goal of science is to extend the range of knowledge, to improve its quality, and to put it to work in the widest areas of human life.

But science is a social institution in a much more significant way: it has become a tremendous force determining where society is going and what it is likely to be in the future. In this respect also science is like other social institutions. Education *usually* works toward the social good. But if it is propaganda and indoctrination it does not. Religion *usually* produces a better society. But we must not forget the Spanish Inquisition, the Holy Wars, and other crimes committed in the name of religion. We must now look at science in the same way. It is trying to make more and better scientists, and therefore a better society. But in the invention of the atom bomb it has diverted much science away from basic research, has increased international tension, and has provided the means for committing racial suicide.

The issue which is involved here is fundamental to the understanding of science. It is usually formulated in terms of the differ-

[1] Bernard Barber, *Science and the Social Order* (Chicago: The Free Press of Glencoe, 1952), Chaps. III, IV, V.

ence between *pure* science and *applied* science, or the difference between *free* research and *channeled* research.[2]

2. Pure and Applied Science

The science which we have been considering thus far in this book has been almost exclusively pure science. The method which we examined in the early chapters was that of theoretical science, not that of technology. The reason we found it possible to proceed in this way is that applied science must rest on pure science. Pure science gives us knowledge, applied science uses this knowledge for the satisfaction of other desires. In free research the scientist is at liberty to follow his problems where they lead him; in channeled research he is restricted by forces imposed from outside. Francis Bacon said that nature to be commanded must first be obeyed. In pure science the scientist obeys nature; he simply learns what nature has to tell him. In applied science he commands nature; he takes what he has learned and employs it to transform nature into something which is more to our liking—a nature which contains more health and longer life, less extreme heat and extreme cold, more food, more rapid and more comfortable travel, less sin and vice, less poverty and war, less fear, and so on through the gamut of the possible positive and negative values of life.

Many have argued that only *pure* science is "good" science; science becomes defiled by being applied. Science, they would claim, has only one obligation—the obligation to increase knowledge of the world. But even this obligation has no "external" claim; scientists pursue knowledge simply because they happen to want to engage in this particular activity. Science can therefore be considered as a game, played by those who happen to find it to their liking. It is similar to working crossword puzzles or reading detective stories except that its problems are set by nature and not by man. This is the conception of science which is presumably in the mind of the common man when he calls the scientist an egghead or a highbrow. He thinks of science as a very remote and academic affair, carried on by a few highly gifted and trained individuals who withdraw from the current of society and retire

[2] See above, p. 172.

into cloistered educational institutions in order to spend their time in utterly useless research.

Such an attitude as this is hard to understand until we recall that in the early years of its development science was looked upon not as a method for earning a living but as the avocation of the cultured gentleman. Voltaire and Mme. du Châtelet used to conduct scientific experiments at house parties for the entertainment of their guests. In the last century it was not unusual for a man of high social standing and great wealth to set up a laboratory in a spare room in his home where, surprisingly enough, he sometimes made important scientific discoveries. Benjamin Franklin pursued science only as an avocation, yet he made significant contributions to physics. Today this is no longer possible because science demands a long period of training and requires equipment which is prohibitive in cost. But the attitude toward science which it engendered remains in many circles. Science is a harmless pastime, entertaining and absorbing, but having no relation to social problems. It may even be looked upon as an escape from society and, like alcohol, as something in which to indulge when the problems of living become too unpleasant to endure. But it should not be condemned or abolished, for it is enjoyed by many people and it is at least a remnant of an earlier culture in which it functioned to embellish life and give it more sparkle.

This extreme attitude toward science, as might be expected, is rare among scientists themselves. While there are many who would defend pure science, few would undertake to do so simply on the grounds that it is amusing to those who pursue it; to deny that science is wholly practical is not to insist that science is a harmless pastime. Science *does* have duties to society, but not wholly in the invention of devices to increase comfort or improve health. A recent writer suggests that scientists should feel "a little cold, mean and selfish"[3] in "piling up discoveries in regions isolated from the general body of knowledge and social welfare."[4] But while the pure scientist may pile up discoveries isolated from social welfare, he does not do this in isolation from the general body of knowledge. The sole responsibility of the pure scientist to society is to learn as much pure science as possible. His goal is

[3] J. G. Crowther, *The Social Relations of Science* (New York: The Macmillan Company, 1941), p. 522.

[4] *Ibid.*, p. 352.

to extend knowledge, not to busy himself with its uses. Knowledge is power, but only when man seeks it as knowledge, not as power. Pasteur considered the phrase "applied science," used apart from pure science, to be a "most improper expression." [5] The pure scientist believes that when he thinks only of control and not of the knowledge which provides this control, he wastes effort; he tries to solve practical problems without having at hand the knowledge by which he can do this. Better, therefore, to pursue truth in its purity and let the applications take care of themselves. Schiller said that for some people knowledge is a goddess; for others she is only a cow, milkable every day. Pure scientists feel that science is a goddess, and they are happy to have those who think of her as a cow take over the milking and become engineers, technicians, and artisans.[6]

3. Science As Social Engineering

Sharply contrasted with this view of science is that which identifies it with what Auguste Comte called "social engineering." According to this conception all science is applied science; science exists only for the purpose of improving life. It provides us with an effective means for satisfying our desires. If we wish health, science will tell us, within the limitations of current medical knowledge, how best to achieve it; if we wish comfort, science will provide us with the latest in mattresses, sleeping pills, and air-conditioned living; if we wish to increase social values, science will provide us with superior housing, labor-saving devices, and effective communication; even if we wish to realize one of the so-called higher values, e.g., beauty, science will make available such indispensable instruments as color photography, paperback classics, radio, television, and stereophonic recordings. In times of war this conception of science is exemplified in an ideal manner. All efforts of science are directed to the winning of the war. Science must make the individual soldier into the most effective fighting unit possible—by providing him with powerful arms

[5] I. B. Cohen, *Science, Servant of Man* (Boston: Little, Brown & Co., 1948), p. 56.

[6] The following story, whose origin I have been unable to locate, illustrates the point: A member of the English Parliament visited Faraday's laboratory and was shown a small model of an electric toy used to demonstrate the magnetic effect of an electric current, and asked, "But what use has it?" Faraday replied, "Some day you will be able to tax it."

and munitions, proper clothing and food, protection against disease, and even suitable amusement for his idle hours.

It seems clear, therefore, that as a matter of fact science does play an important practical role. When men have made known their desires and interests, and have asked science to provide the means by which these could be satisfied, it has responded generously. In general no one seems to have questioned either the right of society to call upon science in this manner or the responsibility of science to meet these demands to the limit of its ability. But the further implications of this situation have not always been seen, and it is only in recent years that we have recognized the seriousness of the problem. If the scientist is to play this practical role must he then assume responsibility for the way in which his discoveries are to be used? Is he permitted, for example, to withhold from the public any information which might be used for harmful ends? And is he to be blamed if information which he does provide, and which seemed to promise only beneficial results, is employed by someone for nefarious ends?

There are two sharply divided schools of thought on this issue. According to one, science should be wholly neutral and should raise no question concerning the *value* of the desires it endeavors to satisfy. Science has nothing to do with distinctions between "good" desires and "bad" ones, between desires which ought to be satisfied and those which ought not, between interests beneficial to man and interests directed toward his destruction.

> No science tells us *what to do* with the knowledge that constitutes the science. Science only provides a car and a chauffeur for us. It does not directly, as science, tell us where to drive. The car and the chauffeur will take us into the ditch, over the precipice, against a stone wall, or into the highlands of age-long human aspirations with equal efficiency. If we agree as to where we want to go and tell the driver our goal, he should be able to take us there by one of a number of possible routes the costs and conditions of each of which the scientist should be able to explain to us.[7]

In the past science has been as ready to turn its efforts to the invention of gunpowder as to the creation of medicines, to the concoction of poisons as to the advancement of antiseptic surgery, to the invention of the electric chair as to the development of

[7] G. A. Lundberg, *Can Science Save Us?* (New York: Longmans, Green & Company, 1947), p. 31. Courtesy of David McKay Company, Inc.

penicillin. When science provides for the satisfaction of a wider range of desires, and for a more effective satisfaction of all desires, it makes possible not only a more abundant life for the good man but also a more completely evil life for the man bent on destruction. The doctor who seeks the health of his patients has at hand an almost limitless range of drugs and operative techniques by which he may accomplish his end; but the power-mad aggressor who wishes to destroy an enemy state has an equally wide assortment of instruments by which he may reach his goal. Research in advertising techniques has effectively shown how to provide the prospective buyer with information on the basis of which he may make rational choice among competing products; but it has also shown how he may be unconsciously persuaded, through packaging devices, arrangement of displays, and constant reiteration of phrases and tunes on television programs, to purchase a product for which he has no real need. In either case science is simply playing a practical role, but *which* it plays is no concern of the scientist. A poll of scientists taken by *Fortune Magazine* in 1946 showed that 79 per cent of them would never withhold a discovery from the world even though convinced that it would be productive of more evil than good.[8]

Furthermore, argues this group, the scientist cannot possibly foresee all the consequences which will follow when the results of his investigations are applied to social ends. One of the important difficulties which is involved in every use of means to achieve ends is that in the satisfaction of certain desires others must be thwarted. The ends determine the selection of the means; but the means, once found, frequently prove responsible for results which were not anticipated. The doctor often finds that the only medicine which will cure a patient is one to which the patient is allergic; the home gardener sometimes finds that the poison which is most effective in killing the weeds in his lawn kills the grass as well; designing engineers, working on the problem of nuclear fission for rail transportation, are discovering that the excessive weight which must be carried in order to afford adequate protection against radioactive bombardment is making the whole project impractical. In general, therefore, when science attempts to put its finger into social processes for the purpose of eliminating evils

[8] Barber, p. 210.

and dissatisfactions it often finds that it succeeds only in multiplying them. In its attempt to realize certain ends it defeats others. Mere mention may be made of some of the unfortunate consequences of the rapid development of science: the increased unemployment due to the introduction of automation into industry, the high obsolescence charges which are the consequence of the rapid changes in the design of machinery,[9] and the general economic instability resulting from a technological society—which has led to the banker's definition of science as that which makes securities insecure. All these factors have led many to ask seriously whether the days before science, the "good old days," the days of the "simple life," were not better. To be sure, these were the days when we could satisfy fewer desires but in the satisfaction of the limited range of our interests we had to accommodate ourselves to a smaller number of dissatisfactions. In any case, however, such problems as what constitutes a good life, and whether one form of life is better than another are not scientific problems, and the scientist has no concern with them.

On the other hand, there is a large group of individuals who feel that the scientist is definitely responsible for the uses to which his discoveries are put. There is a great variation in the way in which this responsibility is understood. At the one extreme are those who unstintingly blame the scientist for practically all of the difficulties in which the world finds itself today. Most serious of all is the cold war which now engulfs us and is the direct result of the invention of the atomic bomb. But this was only the culmination of a long series of inventions for evil—gunpowder, poison gas, flame throwers, killing drugs, and instruments of torture. The biologist has even been blamed for disclosing that in the animal world survival is determined by the "law of the tooth and the nail"; for this has obviously produced the methods of cut-throat competition so common to the business world today. It is not only for the creation of these instruments of evil intent that the scientist is blamed, but for his willingness to serve as a governmental adviser, and often recommender, when decision as to their use

[9] "Broadly speaking, there has not been a time during the past fifty years when anything manufactured by the General Electric Company was not, to some extent at least, obsolete by the time it was put in service." Statement by Owen Young at the 50th Anniversary Meeting of the General Electric Company. Quoted from R. Calder, *Science in our Lives* (New York: Signet Key Books; New American Library of World Literature, Inc., 1954), p. 137.

must be made by the civil authorities. The great increase in the number of technical consultants employed by our government has been one of the most noteworthy of administrative changes which has occurred in our political affairs of the last twenty-five years. This extreme view of the powers and the responsibilities of science led the late Victorian novelist, George Gissing, to write that he hated and feared science because he was convinced that in the future it would be the "remorseless enemy of mankind," would destroy "all simplicity and gentleness of life" and all beauty; that it would "restore barbarism under the mask of civilization," darken the minds of men and harden their hearts, and ultimately plunge the world into "blood-drenched chaos."[10]

Others argue that the responsibility of the scientist requires him to withhold from the public only information which could possibly be used for harmful ends. The free dissemination of information about contraceptives is a case in point; whether this would be generally productive of more happiness than unhappiness is hard to tell. The foresight and vision required to make a sound judgment on such a question is rare among men, and one can readily understand the reticence of many scientists to assume responsibilities on matters of this kind. Others insist that while the *information* should be made public the *uses* of the information should be controlled by law. We all know about tranquilizers, for example, but their use is forbidden except under authorization by a registered physician.

Still others, taking a more moderate position, contend that the scientist has responsibilities for the applications of his discoveries, but only as a citizen, not as a scientist. The Bishop of Ripon recently advocated a "moratorium on inventions"[11] until the social sciences have caught up with the natural sciences and we are in a position to know how to use the inventions which we already have. Even if the technical knowledge of the scientist puts him in a better position to foresee the possible uses of scientific information—very often not the case—he is under no obligation *as a scientist,* according to this point of view, either to release or withhold such information. As a member of a democratic society, interested in the social good, he may feel differently, but in this

[10] Morris Goran, "The Literati Revolt Against Science," *Philosophy of Science,* VII, No. 3, 379. (Quoted from the late Victorian novelist George Gissing.)

[11] Barber, p. 215.

capacity every individual is to some degree responsible for the unforeseen consequences of his acts.

4. Pure Science, Applied Science, and Occupations

I am convinced that much of the confusion concerning the role of the scientist in society is based on a failure to examine more carefully the important distinction between pure and applied science. It seems clear enough on the surface, but the term "applied science" covers an ambiguity. It may mean the *study* of the ways in which nature may be modified in order to produce a better world, and it may mean the *performance* of the acts by which this transformation is brought about. Warren Weaver has an apparently clear-cut statement of the distinction.

> It is specially important to distinguish between *research* . . . and *development*. The two are often combined in both organizational and budgetary statements, as the familiar initials "R & D" illustrate. By looking at the resources devoted to "R & D," one may get a dangerously incorrect impression of what is being allocated to "R." The distinction has been defined by the National Science Foundation in saying that *research* is the "systematic and intensive study directed toward a fuller knowledge of the subject studied," whereas *development* is the "use of that knowledge directed towards the production of useful materials, devices, systems, methods, or processes." [12]

Aristotle insisted that there are three types of science: theoretical, practical, and productive.[13] By "theoretical science" he meant what we have called "pure science." In the introduction of practical sciences he suggested that perhaps there is a sense in which applied sciences are not really productive according to the usual meaning of the word, but merely inform us how we should proceed in case we wish to alter nature in certain ways. If this is the case then applied science is really a science in much the same way that pure science is: it informs us about nature, but the information which it provides is in the form of causal laws stating what means should be employed to produce certain desired ends.

[12] Warren Weaver, A Great Age for Science, from *Goals for Americans* © 1960 by The American Assembly, Columbia University, New York, New York, by permission of Prentice-Hall, Inc., Englewood Cliffs, New Jersey. P. 109.

[13] Aristotle, *Metaphysica*, Chap. I, and *Ethica Nichomica*, Book VI, Chaps. III-VII, *Works of Aristotle Translated Into English*, ed. J. A. Smith and W. D. Ross (Oxford: The Clarendon Press, 1910–31).

It does not state whether these ends are good or bad, nor does it actually modify nature by introducing the required means. But it does provide the "know-how," the theoretical grounding of skills, where this is known, and the "rules of thumb," where the theoretical principles are not known. That such study is closely tied up with the actual modification of nature is unquestionable, but that it can be carried on apart from such manipulation and transformation is clearly indicated by the difference between engineering schools and factories, agricultural colleges and farms, schools of business administration and business enterprises, and medical schools and practicing doctors.

Let us try to clarify this distinction between the practical scientist as a scholar and the practical scientist as one who alters the world in such a way as to produce a desired end, by calling the latter an "artisan," a "practitioner," or a "professional man."[14] I think we can then see that in all spheres of activity there is a distinction which can be drawn, at least theoretically, between pure science, practical science, and a large group of activities including skills, practices, trades, professions, and occupations. "Pure science" can then be defined as a study of causal relations (it deals with truths of the form, "A *causes* B"), and "practical science" can be defined as a study of how these causal relations can be put to work when A is a *means* and B is an *end*. It is an attempt to determine how to produce A if B is a positive value, and how to prevent A if B is a negative value. For example, if B is health, medicine is the practical science which shows how to produce all of the conditions which result in health, and if B is disease, medicine informs us how to avoid all of the conditions which result in illness. On the other hand, the trades and professions, accepting the values set by man and society, and using these means-end relationships, attempt to *create* the means which result in the positive values and try to *prevent from occurring* the means which eventuate in the negative values. In the present example, the practicing physician *pre*scribes a plan of living which will make us healthy and *pro*scribes a regime which will make us ill. An illustrative table of various sciences and occupations is given in Figure 14.

[14] An alternative method for interrelating science, philosophy, technology, and the humanities is given by H. G. Cassidy, "The Muse and the Axiom," *The American Scientist*, Vol. 51, No. 3, 315–326.

The relations between Columns *IV* and *II* exhibit normally a common pattern. In order to be a good applied scientist one must be well informed in the theoretical content of the science; applied science depends on pure science. But the pure scientist, however superior he may be, is not necessarily a good practical scientist, for in taking over the responsibilities of the latter he must acquire familiarity with ends and goals. To be a good applied scientist requires that one have a grasp of the uses to which the science can be put. But the recognition of these uses does not, I believe, require any commitment on the part of the applied scientist, *as a scientist,* as to the desirability of these goals. As an applied scien-

FIGURE 14

TABLE OF PURE SCIENCES, APPLIED SCIENCES, AND TRADES AND PROFESSIONS

II PURE SCIENCES (the sciences and critical metaphysics)	IV APPLIED SCIENCES (study of the use of pure science to attain goals)	V TRADES AND PROFESSIONS (use of applied science to attain goals)
A. Pure Behavioral Sciences	A. Applied Behavioral Sciences	A. Preferential Behavior as Life-Work
Political Science	Science of Political Behavior	Politics
Sociology	Science of Social Work	Social Work
Economics	Science of Business Administration	Commerce, Industry
Science and History of Religion	Science of Ministerial Training	Preaching
Descriptive Morals	Science of Character Building	Moral Reforming
Description of Art Objects and Experiences	Science of Art Skills	Artistic Creation and Criticism
Human Physiology and Psychology	Medical Science	Medical Practice
B. Pure Biological Science	B. Agricultural Science	B. Agriculture
C. Pure Physical Science	C. Engineering Science	C. Engineering, Manufacturing
D. Logico-Mathematics	D. Applied Mathematics and Logic	D. Computer Operation (?)
E. Critical Metaphysics	E.	E. ....

tist he does not determine the direction which his science should take in its forward movement. The goals are presented to him by the particular society in which he lives—the cultural milieu of which he is a part. Of course if he is a loyal member of this society, and it is a democratic society, he contributes to the decisions as to what its goals shall be. But as a scientist he need neither accept nor reject these ends; he is assigned them by society and his task is to show, to the best of his knowledge, how they can be achieved. This relieves him of all blame, as a scientist, when in response to the demands of the military men he shows them how to produce a powerful weapon of war. And it also wins for him no praise when he creates a method for producing a serum against polio. In both cases he is simply doing his duty in meeting the demands which society imposes on him, and his reward lies in having done his job well. In the same vein he is relieved of responsibility when a tool which he has designed to produce a desirable end proves to have unanticipated evil uses, and he can ask for no praise when he shows how to produce an instrument for the destruction of man, and this proves to have unforeseen benefits. No scientist can foresee all of the effects of a given cause, for society is far too complex and the ultimate future is largely unpredictable. Was Euclid to blame for the atomic bomb because geometry was used in its construction?

The lines between these columns cannot, of course, be sharply drawn. Consider again Columns *IV* and *II*. If for the moment we think not of the sciences but of the *teachers* of the sciences we can see how hard it would be to distinguish one who teaches, say, pure political science from one who teaches applied political science; perhaps the former would place greater stress on the structure of government while the latter would tend to emphasize the ways in which politicians operate in attempting to accomplish their ends. Similarly a sociologist who teaches pure sociology probably would concern himself more with the study of social organization, and would leave to the teacher of social work the study of how most effectively to bring about desirable social changes, especially in the case of specific individuals or groups. Perhaps the distinction would be sharper in the case of pure human physiology and psychology, on the one hand, and medical science, on the other; pure human physiology and psychology would simply accept the facts that there are malfunctionings, deformities, and death, as well as health and well-being, while the medical scientist would be much concerned with how to prevent the former and promote the latter.

The lines between Columns *IV* and *V* have already been shown to be relatively sharp. While the notion of "trades and occupations" must be left somewhat vague, since it is designed to include hobbies and avocations as well as occupations by which an individual maintains himself in society, the distinction between the scientist who knows how to produce certain kinds of goal and the artisan or professional man who puts this knowledge to work is not hard to make. Furthermore, some more or less artificial occupations are bound to appear in the effort to complete the diagram: I do not know of any profession of "moral reforming," apart from preaching or social work, for which we would receive training by a "scientist of character building." (I am sure that all of us can discover, even among our friends, certain individuals who are "do-gooders," especially if this avocation in which they engage is directed toward ourselves.) But that there *is* a science of character building is clear from the fact that when a moral delinquent appears on the scene the responsibility for his condition is usually denied, depending upon who the "denier" is, by the home, the school, the church, the police, the legal profession, the communication media—moving pictures, television, radio, bookselling, and the rest. Perhaps it is just because of the plurality of "applied sciences of character building" that there are no moral reformers. But one can hardly deny the need for them.

If we now combine Figure 11 [15] and Figure 14, we can see how complicated the social role of science can be. If we limit ourselves to the modes of valuational behavior—political, social, religious, moral, aesthetic, and the rest—we can see that with each of these there is correlated *four* other social manifestations.[16] Corresponding to political behavior there is political science, philosophy of political science, applied science of political behavior, and politics. Corresponding to aesthetic behavior there is the descriptive science of art objects and experiences, philosophy of art (aesthetics), applied science of developing art skills, and artistic creation itself. That these five aspects of human living intermingle

[15] See above, p. 186.

[16] Karl W. Deutsch, "Scientific and Humanistic Knowledge in the Growth of Civilization," *Science and the Creative Spirit*, ed. Harcourt Brown (Toronto: University of Toronto Press, 1958), Chap. I.

almost inextricably is obvious. But that they can in many cases be distinguished, though perhaps not separated, is an important fact — important for curricular planning in schools and colleges, important for determining the scientist's social and political responsibilities, and important, as we shall see in our final chapter, for determining one's design for living.

Again combining Figures 11 and 14, the relation between Columns *V* and *IV* is substantially the same as that between Columns *IV* and *II*. Skills, at least the higher skills, are dependent on applied science. To be sure, there are many men who become artisans and even skilled workers without knowledge of principles; they arrive at their position through the school of hard knocks. And, as above, there seems to be something required of the workers and professional men which is over and beyond the competence and interest of the applied scientists. Not all who know how to build good bridges can do so, and not all who know the principles of social work are successful social workers. Whether those who are thus lacking in certain skills are suffering from a deprivation is a matter of point of view. George Bernard Shaw stated once that those who can, do; while those who can't, teach. This fact is also responsible for the use of certain derogatory words, so frequently alluded to in the preceding pages, to characterize scientists. We can now see that this terminology may be applied to the practical scientist as well as to the pure scientist; as an applied scientist he may know how to accomplish certain ends, but as a worker or laborer he may be unaware of some of the important tricks of the trade and might bungle a job which is assigned to him. The problem is frequently formulated in terms of the relative value of *knowledge* and *experience* as ground for a successful pursuit of the trades, professions, and arts. Which is the better preparation for becoming a good farmer — years on a farm or a four-year course in agriculture at a university? Which is the better preparation for becoming a good politician — rising through the ranks from city councilman to United States Senator or teaching political science in an institution of higher learning? Which is the better preparation for becoming a successful businessman — a long period in the business world or a strong training in the economics of marketing and finance? Of course the situation rarely involves this sharp dichotomy, but when it does the decision is not easy to make.

Attention should be called to items *B*, *C*, *D*, and *E*, in Figure 14. Since the subject matter of the pure sciences on these levels is not human behavior, but life, mass-energy, abstract structure, and the categories, the social relevance of these disciplines is not so close. In each case (with the exception of critical metaphysics) there is a pure scientific study, a philosophical study, an applied study, and usually large classes of trades and occupations. The high degree of generality involved in critical metaphysics is responsible for our inability to separate the scientific aspects from the philosophical, and the pure aspects from the applied. So far as I know the only way to "practice" metaphysics is to teach it. One frequently sees signs, "Practicing Metaphysician," on office doors; but if one is rash enough to invest in the cost of a visit to such an authority he is sure to be rewarded by receiving a mixture of astrology, theosophy, phrenology, and fortune-telling. To introduce the metaphysician as a teacher calls attention to a fact not heretofore emphasized in our discussion. Teaching is itself an occupation, and therefore will have its four correlates. Education is one of the important value experiences, whether viewed from the perspective of the student or that of the teacher. But if there is education, there is a science of education, a philosophy of education, an applied study of the means by which knowledge can be most effectively transmitted from teacher to student, and the teaching profession itself.

As a result of these considerations I think we can see exactly where we stand on the question of the social responsibility of the scientist. The pure scientist has only one duty—to increase our knowledge of the world in accuracy, certainty, and scope. In this task no door is closed to him, except when his experimental subjects are human beings and, with some restrictions, the higher animals. (This is not the place to go into the intricacies of the interpretation of the Hippocratic oath or of the justifiability of the claims of the Society for the Prevention of Cruelty to Animals.) With these exceptions no areas are out of bounds for exploration by the pure scientist, and no charge of being anti-social, or even unsocial, can be brought against him if he chooses to explore realms which seem to have no relation to the problems of practical living.

The applied scientist, as opposed to the artisan and practitioner, has two duties. One is that of discovering how the knowledge produced by the theoretician can be put to practical use. But he is not, as an applied scientist, responsible either for the discovery of the practical ends or for the evaluation of them as desirable or undesirable. Society does this for him. Since he is a member of this society he is *indirectly* responsible both for the discovery of these ends and for their evaluation. But it is not in his role as a scientist that he does this, for every member of the society has this same responsibility; we all determine in subtle and indirect ways what the social values are to be and whether they are ultimately worth striving for. But the applied scientist has another important duty. As a scientist he has the responsibility to inform the public, and especially that portion of the elected officials who have to make major policy decisions, concerning the nature of the instruments which science is about to propose to achieve certain social ends. This means that he must seriously attempt to translate his technical language into words understandable by the non-specialist in order that the latter may make informed decisions concerning the use of the instrument. It does not mean that all governmental information should be declassified, or that every manufacturer who is engaged in research should be required to tell all of his competitors what lines of investigation he is pursuing. But the information asked of the scientist, it seems to me, should be restricted to technical information concerning the product itself, and should not extend to opinions as to the desirability of its use. Here the scientist has no more foundation for judgment than does the businessman, the farmer, or even the ordinary citizen.

The artisan and the practitioner, as opposed to the applied scientist, has a still different task. His job is to realize these values by actually transforming the world — by manufacturing home appliances and by building roads and bridges, by curing illnesses, by increasing our wealth, by improving our government and strengthening our nation, by raising our standards of living, by increasing our food supply, by providing us with books and objects of aesthetic enjoyment, by strengthening our religious faith, and by increasing the many other values which our complex society is in a position to offer. Here commitment to values does seem to be involved in a way which is lacking in the case of the applied scientist. For no one would presumably spend his life in the attempted realization of a certain value unless he were convinced that it had something to contribute to man's happiness and well-being. A significant mark of a democratic society is the opportunity it offers to every man, at least in theory, to choose the area within which he may pursue his own goals.

5. Science and One World

Another effect of science on society is not so obvious. Science, perhaps more than anything else, is responsible for the "one world" in which we live today. Our highly integrated and organic society is almost wholly the result of the applications of science to the problems of communication and travel. The world is not only smaller than it has ever been before, its parts are more intricately related to one another. What happens in one area of the globe has instant repercussions everywhere else. A steel strike in the United States upsets the world market; a revolution in Africa causes tensions on other continents; a purely local disaster, such as a flood or earthquake, arouses sympathetic feelings in remote areas and often induces utter strangers to sacrifice money, food, and clothing to help the unfortunate victims. Society today is very much like a human being—a disease in one locality makes it "feel sick all over." Isolation, national and personal, has disappeared. The days of the old agricultural society are gone forever—days when the farmer was out of touch with the rest of the world for weeks at a time, when a man often spent his entire life, from birth to death, without going more than a few miles from his home, and when happenings in remote areas of the world were as unreal as events in story books.

Science has therefore, in effect, created the United Nations. That this was not foreseen by science again serves to illustrate the impossibility of forecasting the uses to which science will be put. Unfortunately the United Nations has not solved the problem of cultural conflicts. I suppose no one would blame science for this. But science is in the happy position of being able to make its contributions to this problem. There is bound to emerge, in the not-too-distant future—barring extermination of the human race by nuclear war—an intercultural morality, which must be binding on all nations and all men. Since science, as we have already seen, is not concerned with values, the creation of this morality cannot be its task. But it can provide us with some of the instruments which are essential to the realization of this goal—increased literacy, more widespread education, cheaper and more effective mass communication, including intercontinental television, improved methods for teaching foreign languages and for translating from one language into another, and increased opportunities for cultural exchange between the nations of the world. Only by enlarging the social consciousness of the individual and by developing in him a realization of his responsibilities to the larger group of which he is becoming a more and more important part can this be accomplished. Science cannot do this job, but it can provide many of the technological instruments which are indispensable to its success and which, granting the desire of peoples to understand one another, can be used to speed up the realization of the goal.

6. The Democracy of Science

Such considerations as these remind us that science has another important role in society. The salvation of the human race in a highly integrated world rests upon an increased social consciousness, which seems to be most effectively fostered by the democratic way of life, with its freedom of thought and speech, and its provisions for solving its problems through cooperation among men of diverse interests and capacities. Science once proudly offered itself as an outstanding example of the effectiveness of cooperative effort. Until recently science, like art, music, literature, philosophy, and religion, had never known national boundaries; there was no "Jewish science" or "Catholic science," no "Caucasian science" or "Negroid science." Furthermore, it had always been willing—indeed, it had always felt a strong obligation—to make its discoveries public at the first opportunity. From the time when it broke its alliance with magic and necromancy it realized the futility of secrecy. Faraday invented a formula—*work, finish, publish*—which he considered to be the basis of all success in science. This feeling of responsibility on the part of the scientist to communicate his results to others is coordinate with his realization that his own results are in turn dependent on information provided by his colleagues. We saw in an earlier chapter that one of the most important methods for acquiring data is the acceptance of "reports" from other scientists working on the same problem. Science was characterized, therefore, until recently, by the friendly and cooperative spirit which was proudly exhibited by its members; scientific notions were public property and were there for whoever wished to use them in whatever manner possible. Weaver writes, "The television critic, John Crosby, writing of some science shows in the *New York Herald Tribune* of March 30, 1960, declared: 'There is something so ennobling, so uplifting, watching men wrestle with the mysteries of nature rather than punch each other in the nose.'"[17] As recently as twenty-five years ago a committee of the American Association for the Advancement of Science announced its opinion that it was "unethical for scientists to patent the results of their work."[18]

Unfortunately when pure science became applied science and applied science became either industrialized or militarized science, it could no longer point to itself as a shining example of the democratic process at work.[19] Cooperation among industrial engineers has not been noteworthy in recent years. In fact a competitive economy forbids such cooperation; rivalry, with its necessary secrecy, is the approved attitude. There are reasons, of course, why this must be the case. The manufacturer is motivated by conflicting aims: he must make the best product and he must sell as much of it as possible. In order to achieve the first goal he must learn from other industrial engineers how they are meeting their problems. But in order to achieve the other goal he must discover some scientific truth, unknown to his rivals, which will enable him to produce something of higher quality, of greater durability, or at a cheaper rate, than his competitors can; he can out-sell them only if his manufacturing formulas are not made public. The resolution of this conflict in the minds of most industrial engineers is not difficult: the company for which they work makes up their minds for them, and secrecy usually prevails over cooperative activity.

Much the same sort of thing happened when applied science became a tool for national survival. One can hardly say that the

[17] Warren Weaver, A Great Age for Science, from *Goals for Americans* © 1960 by The American Assembly, Columbia University, New York, New York, by permission of Prentice-Hall, Inc., Englewood Cliffs, New Jersey. P. 105.

[18] Barber, p. 153.

[19] J. Bronowski, *Science and Human Values* (New York: Harper and Brothers, 1951), Chap. 1.

study of rockets and missiles has been characterized by international cooperation. Indeed, it has created something which was almost unthinkable fifty years ago—a *geography of science*. While there has always been such a study (since a scientist must always be *somewhere* when he carries on his investigations) it lacks so many of the characteristics commonly associated with science that one feels tempted to deny it the name. Space science today cannot be called "science" but only "Russian science" or "American science" since it is the product of cooperation only in the sense that Russians cooperated with one another, and Americans cooperated with Americans. While it is true that this new "national" science is applied science rather than pure science, it has affected the latter in significant ways. For the impossibility of foretelling what consequences pure science may have in the development of missiles and other instruments has required that this knowledge also be classified. Furthermore, pure science tends to be discouraged, since the urgency of the cold war problem demands that science be put to work immediately, and only in those areas where national security is endangered. This is surely science in a strange garb.

The fact seems to be that science, once an outstanding example of the democratic process at work, has now become the instrument of an undemocratic, competitive society. When rival cultures struggle for supremacy they must use whatever tools they have available to achieve victory. And when industries gamble for markets they cannot lay their cards on the table and expect to maintain their superiority over their competitors. When science becomes the instrument of society then society can, in a measure, dictate what science is to be and do. The result is that science undergoes significant changes: cooperation ceases, the urge for carrying on basic research weakens, and secrecy takes over.

There are many signs that international cooperation among scientists is on the increase. The International Geophysical Year was eminently successful, both in the spirit in which it was carried on and in the results accomplished. Cooperation among scientists outside the iron curtain proceeds much as it did before the cold war. But even as I write preliminary conversations have taken place with the Russians on a joint project for exploration of outer space with a view to improving the accuracy of weather predictions and extending facilities for intercontinental communications. Nuclear testing in space has been banned. Cooperation among medical scientists on problems of world health would seem to lie just around the corner. Only the future can disclose whether science is ever to resume its position as the supreme example of an enterprise which knows no national boundaries.

7. The Preservation of Basic Research

The problem of saving science today is essentially one of preserving and encouraging basic research. Perhaps the most important factor in bringing this about is dispelling the notion that pure science is, on the one hand, poorly defined and, on the other, useless. Let us examine the former of these misconceptions. The business executive who defined basic research as "when you don't know what you are doing" expressed an opinion which is widely held. But nothing could be farther from the truth. While it is true that in basic research the scientist is, in a sense, "searching for theories rather than facts" and thus appears to be pursuing something very elusive, he is, presumably, confused neither about the theories he is searching for nor the facts which lead him to make the search. In this sense he clearly knows what he is doing. Possibly he may find it harder to discover theories than to create facts, since, as we saw in the early chapters, there are no rules for getting theories. But neither are there rules for getting facts, especially in the absence of theories to guide us. No doubt many scientists who engage in pure research do not clearly define their problems at the outset, and thus waste time in ineffective thinking. But there are bad workers in any human endeavor, and one should not measure the endeavor by its bad examples.

But neither can one rightfully say that pure science is useless. One can, of course, *define* basic research as impractical science, but then the issue becomes one of words only. To put too much stress on use results in absurdities. When Benjamin Franklin was asked whether every new fact should not be measured in terms of its usefulness he replied, "Of what use is an infant?" We should certainly not be misled by the word "impractical" which may mean either *useless* or *not at present known to have any use*. The former is a derogatory term and suggests wastefulness or, at best, sterility. No one, I believe, who knows anything about basic research, would be willing to attribute to it either of these characteristics, for no one without infinite wisdom can predict what future applications any scientific theory may have. The history of science clearly shows us that theory after theory which was claimed (sometimes boastfully) to have no practical applications, has been later put to work in totally unforeseen ways. The famous mathematician who recently declared that he would not publish in the future any results which "might do damage in the hands of irresponsible militarists"[20] certainly did not understand the role of science. To correct this misconception about basic research, therefore, we have only to insist on the triple fact that it is not confused or poorly defined, that all pure science, so far as we can tell, eventually becomes applied science, and that no applied science is possible without previous pure science. This would seem to justify at least some attention to pure science.

The primary problem in the continuation and growth of pure research is that of increasing the rewards which are offered to those who are engaged in it, and to those who are interested in entering it. While a sincere devotion to science is a presupposition to success in the field, and probably will induce many of those presently pursuing it to continue to do so and may even attract many neophytes, nevertheless such "fringe benefits" as better salary, better laboratory equipment, improved working conditions, and freedom to select one's problems and solve them as he pleases must be considered. Since all these reduce, in the final analysis, to matters of financial support, the problem may be formulated in this way.

There are four main sources for the support of basic science—education, industry, the private foundations, and the national government. Of the total money spent in the United States for research and development in science approximately 98 per cent is provided by federal and state governments and industry, only 2 per cent by colleges, universities, and other non-profit-making institutions. (These figures are, of course, only rough estimates and do not include data covering the last two or three years.) But government and industry allocate to basic research only about 7 per cent of the total amount provided for science, while education and the foundations assign to pure research as much as 67 per

[20] I. B. Cohen, *Science, Servant of Man* (Boston: Little, Brown & Co., 1948), p. 294.

cent. Weaver, as we saw earlier,[21] points out that these figures are quite unreliable because many industries which have research and development accounts make no distinction between the "R" and the "D," and because, even where the distinction is made, the principle on which it is based varies considerably in interpretation from industry to industry. But the figures lead to the conclusion that of the money spent in our country in support of science approximately 8 per cent is available for basic or fundamental research. While the total money spent for aid to science is bound to increase in the future, there are equalizing factors in both pure and applied science which tend to keep the proportions more or less constant.

When we consider separately the various sources for the financial support of science I think we can readily see that the schools and colleges are seriously restricted in the contributions they can make. Their primary job is to educate, and whatever one may say about the intangible benefits accruing to teachers who are also scholars, and to colleges and universities which employ teachers who are also scholars, the bitter fact remains that when income is limited, research is more likely to be sacrificed than teaching. With the anticipated rapid increase in enrollment in the next ten years the prospect for more time to be devoted to research is dim indeed. Concessions in the form of sabbatical leaves, occasional research grants, and temporary reductions in teaching loads are, at least in certain fields of study, so infrequent as to be hardly worth mentioning. Universities and colleges have been in the past the places where great discoveries were made. Up to a few years ago, of the 131 Nobel prizes awarded in physics and chemistry 116 had gone to teachers and only 15 to industrial scientists. This proportion will almost certainly change in the future as more and more of our best scientists are drawn away from our academic institutions by the higher salaries offered in industry.

Industry is in a much better position to support scientific research, largely because it has more money. And because industry can offer more support its frequent raids on the scientific departments of our colleges and universities are, almost without exception, successful. Industry's affluence also draws graduating scientists away from careers in teaching and research into careers

[21] See above, p. 247.

in industry. Whether pure science gains is another matter. Usually it does not, for reasons which we have already pointed out. But the situation seems to be changing for the better. Much of the research now supported by Remington Rand, United States Steel, and General Electric—to mention only a few—seems more properly characterized as industrially *sponsored* than as industrially *directed.* While industry usually claims priority in the right to use information disclosed by such research, increasing freedom is being given to the scientist in his choice both of the problems which he studies and the methods which he uses in solving them.

The third support for pure research is the private foundations—Ford, Rockefeller, Guggenheim, Carnegie. The importance of this kind of aid cannot be over-emphasized. While grants from these sources are necessarily limited both in amount and in duration, they are commonly not restricted in subject matter, and therefore provide excellent opportunity for basic research which is not hedged in by considerations of utility and practicality. Indeed, the importance of this form of sponsorship lies precisely in the fact that the support is not given subject to the condition that it produce *any* results. Properly speaking, the aid is for the *process* of research, not for the *results.* Basic research as an activity is a fact; basic research as the production of consequences, either in the form of greater wisdom or of better living, is wholly conditional. Goals are often not achieved simply because the problem proves to be insoluble or much more complicated than was anticipated, or because the problem rests on another problem which has to be solved first, or because of any one of a number of other surprises which turn up during the investigation. Basic research is, and always will be, a gamble.

The final source for grants in support of research is the national government, which is playing an increasingly important role. Among the many agencies providing assistance are the National Science Foundation, the Department of Defense, the Atomic Energy Commission, the National Defense Education Act (under the Department of Health, Education and Welfare), the Department of Agriculture, the National Institute of Health, and the National Bureau of Standards. The restrictive conditions which are stipulated by these groups vary from agency to agency. Often the acceptable problems are limited to those which have a distinctly practical application; sometimes, as in the case of the National Science Foundation, there are no such restrictions. The blessing which the government provides in offering this aid to basic research is not an unmixed one. Charges of socialism and favoritism are bound to be leveled against it. As in the general case of federal aid to education the problem is to make the grants where they are needed, yet to allow the applicants a wide range of choice in selecting their problems. Certainly the government would not spend its money wisely if it offered support to *any* scientist who wanted to work on *any* problem which was to his liking. Nor would it do well to demand that every problem have practical application to a specific area of national defense, or to a restricted area of our agricultural economy, or to a particular industry which contributes handsomely to our national wealth. When the problem is solved the answer will probably lie midway between these extremes. If the national government, in cooperation with the United Nations, can resolve this conflict in the immediate future it will provide food for an aspect of our culture which is suffering badly from undernourishment.

Science and the Design for Living

It should be abundantly clear as a result of our discussion that science and living are significantly interrelated. Many of these forms of connection have been examined in the preceding pages. (*a*) Since living is the more or less conscious and controlled pursuit of the basic and fundamental human values, of which truth is one, science is a definite part of living. (*b*) As the pursuit of a value science is not sharply different from art, religion, morality, social behavior, work, or play, and shares many characteristics with these other forms of valuational activity. (*c*) Since it is an intellectual pursuit of a certain kind, science is both like philosophy and different from it. (*d*) Being an intellectual pursuit it requires justification in the same way that religion and the other value pursuits do, and its justification (apart from a more sweeping philosophy of life) is no better or no worse than the justification of the others. (*e*) Since science is a cultural enterprise it occupies, or at least should occupy, an important place in the educational scheme. (*f*) As a social activity it both determines our group behavior and is determined by it.

1. Need for a Scale of Values

But science is not to be identified with living. Living is surely more than the *sum total* of these activities, and more than the mere *pursuit* of these values. Happiness is measured not in terms of the number of values sought nor in terms of fruitless striving, but in terms of patterns of values and progressive realization and achievement. Since the greatest impediment to the maximum attainment of values is the fact of conflicts between them, we are prepared for adequate living only when we have provided ourselves with a scale of values on the basis of which we can make a selection in case a sacrifice is required.[1] We can achieve the greatest happiness if the values we have to reject are those which we deem to be the less important ones, and if the values which we are thus enabled to attain are the more important ones. But such a choice requires that we have a "philosophy of life" or a "design for living." Living is more than the mere pursuit of conflicting values; it is the *intelligent* pursuit of such values under the direction of an adequately formulated philosophy of life. The happy life is one in which there is a maximum of wise choices.

The building up and establishing of a philosophy of life involves two main sorts of activity. The first is the process of exposing oneself to values. This might be called "tasting values."[2] It involves trying out the values either personally or vicariously, through literature and the arts, history, and human associations. The goal is to experience them as directly as possible, and to explore them for what they contain, positively and negatively. For example, one of the most common sources of dissatisfaction in the pursuit of values is the disappointment which one feels after he finally achieves what he has been seeking so fervently; after he gets what he wants he doesn't want it. Many who strive for wealth find their lives empty after they succeed in achieving it; many who have long looked forward to retirement are bored and unhappy when they finally attain it; even intellectual values, as we learn from Faust, may prove in the end to be disappointing. This amounts to a conflict between anticipated and realized values, and

[1] DeWitt H. Parker, *Human Values* (New York: Harper and Brothers, 1931), and W. G. Everett, *Moral Values* (London: Wm. Heinemann, Limited, 1920).

[2] For the contrasting terms "tasting" and "testing" (see next paragraph) I am indebted to Herbert W. Schneider, *Morals for Mankind* (Columbia: University of Missouri Press, 1960).

usually occurs because of the failure to understand the value at the outset; the man who is unhappy after he retires simply did not realize that an empty life is an unpleasant one. The safeguard against these shocks is a more sweeping experience of values. The best goals in life are set not blindly but with intelligent foresight.

But there is a second process involved in establishing a philosophy of life. Values must be not only "tasted" but "tested." Variety is the spice of life only if there is some pattern in plurality, only if the values have been evaluated against one another. Setting up such a scale of values requires balancing of values with one another—of present values with future values, of low quality with high quality values, of dependable values with risky values, of values requiring many sacrifices with those requiring few, of brief but intense enjoyments with those which endure but are more mild. The outcome of such a critical appraisal is a value perspective on the basis of which decisions in life can be made—decisions which are relatively satisfying because founded both on the widest range of possible experiences and on the intelligent estimation of the values of those experiences when they are weighed one with another.

Since the personality of man is a unit, the task of setting up a philosophy of life is a complicated one. When we break up his activities into such clusters as art, religion, morality, and inquiry we should be clearly conscious of what we are doing. We are not really cutting up man, for his experience cannot be analyzed into separable parts. We are simply performing an act of intellectual scrutiny; we are isolating these aspects of man's behavior for the purpose of talking about them more effectively. But they must be put back into man from whom they should never have been taken in the first place. Consequently when we say that inquiry is the collection of experiences which are involved in the pursuit of truth, art is those responses directed to the creation of beauty, morality is concerned with right living, and religion is a matter of striving for holiness or spirituality, we should remember that these activities and experiences interact and intermingle in the life of man in profound and complicated ways. Perhaps it would be more accurate to say that they are fused in the life of the *ideal* man; unfortunately for many of us certain of these activities are placed in water-tight compartments and enter into our lives only on special days of the week or certain hours of the day. But in the truly integrated personality such separation is impossible.

Here we need only point out that our lives are happy just to the extent to which there are not too many conflicts between values which we consider important to our well-being. That there are such conflicts is obvious; values simply are sometimes antagonistic and incompatible, and often we cannot realize all of them. Frequently in order to achieve wealth we have to sacrifice health, or in order to realize beauty we have to sacrifice truth. This incompatibility of values in certain types of situation occurs over the total range of preferential judgments from such simple situations as the impossibility of having our cake and eating it, to the more sweeping and more profound conflicts of a Socrates who must decide between an honorable death and a dishonorable life. Such conflicts can be avoided, or resolved when they do occur, only when we have an adequate philosophy of life. This usually takes the form of a value scale in which the many "bests" of life are so arranged as to indicate clearly which is *the* best, and therefore rarely to be sacrificed to any of the others. A value arrangement of this kind provides a pattern for making decisions in the problems of everyday life, for the "higher" values are generally considered to be more significant and when conflicts with the "lower" values arise the latter are to be denied. The happy life is one in which these conflicts have been reduced to a minimum, or one in which the methods for resolving them are so well developed that decisions are more or less automatic. If we have built up for ourselves through reflective examination of life and its problems an adequate value scale, and have trained ourselves to act promptly and intelligently when a value conflict arises, we are in a good position, other things being equal, to live a happy life.

All of this is, of course, an extreme over-simplification of the problem. It neglects entirely the fact that at times a "higher" value may, with justification, be sacrificed to a "lower" one, that there may be a convergent value to which both values contribute, and that our value perspective is not static and fixed but is itself a reflection of our lives and thus subject to change through intellectual growth and advancing years. It also disregards the fact that the individual who tries to perfect his value pattern by selecting a single "top" value to which all others are to be sacrificed usually turns out to be a neurotic, and seldom fits well into the social scheme. In fact we have invented derogatory terms to characterize these individuals—the esthete, who makes a show of his artistic discrimination and will not tolerate ugliness in any form; the pedant or high-brow, who concentrates on intellectual values; the "do-gooder," who thinks only of moral values and is always eager to tell us when we are disobeying the ethical code (which usually turns out to be *his* ethical code); the religious fanatic whose zeal knows no bounds and who considers it his personal task to save all of those who, but for his consideration and influence, would surely end in purgatory; the hypochondriac, who seeks personal health by continually trying new diets, new pills, quack medical practices, and other panaceas; the miser, who seeks only wealth, not to use it but to hoard it; and so on. Such people fail to realize that happiness is not concentration on a single value but on a rational apportionment of the many values which life has to offer. Apart from a design for living all values are on the same level, and only on the basis of such a plan can wise judgments concerning values be made. The problem of the conflict between science and religion, as we saw in Chapter 11, cannot be settled *in* religion or *in* science, for it involves the relative roles of knowledge and faith in one's value scale. This can be settled only by philosophy.

2. Does Science Affect Our Philosophy of Life?

If we realize, then, that the problem of living is essentially a matter of striving for certain goals which we have set up for ourselves either deliberately or largely as the unconscious result of environmental forces playing upon us, and that effective living depends on having a speaking acquaintance with the widest range of values placed on a preferential scale, we are able to ask a somewhat more meaningful question concerning the relation between science and living. Just how does science affect our philosophy of life? Does a scientific society, such as we are living in today, have a characteristic design for living which a non-scientific society does not have? For example, do we have a more adequate philosophy of life than our prescientific ancestors could have had?

And *if* science does influence our value pattern just how does it do so?

The most obvious answer is that science *does* affect our philosophy of life. As we have already seen repeatedly, science has contributed to our manner of living in a very important way—it has made us more comfortable than we have ever been before. We have less noise, less dirt, less intense heat and excessive cold; we can travel faster and with greater ease; we do not have to work either so hard or so long; we have a much higher standard of living; we live longer and in less pain; and we have almost unlimited time for recreation and play. We have therefore adopted a materialistic philosophy of life—materialistic in the sense that we have raised to a pinnacle the so-called "lower" values of material goods, physical comfort, money, and play, while we have tended to degrade the corresponding "higher," spiritual values of wisdom, aesthetic appreciation, religious fervor, and moral goodness. It is a philosophy which the average European, rightly or wrongly, believes to be the value scale of all Americans.

With the question as to whether this *is* the characteristic American philosophy I shall not be concerned. Probably many Europeans who have not visited America are entitled to believe it on the basis of the propagandizing effect of American movies, comic books, popular music, dancing crazes, and even the behavior while abroad of certain American tourists. The question we are considering is whether science has produced this materialistic philosophy, granting that it does exist. Obviously there is a sense in which science has done exactly this. Physical science has increased gadgets and made daily living more pleasant; medical science has improved health and made us physically more comfortable; all of the sciences have made it easier to accumulate money and easier to spend our earnings in the production of superficial pleasures.

But science has not *urged* this life upon us, nor guaranteed that it will be creative of greater happiness. For, on the one hand, none of these scientific accomplishments has proved to be an unmixed blessing; it has been productive of evil as well as of good. And, on the other, science has never assured us that when we become more comfortable we will necessarily become happier. We do not find that when we travel at fifty miles an hour we are twice as happy as when we traveled at only twenty-five; nor do we find that we are twice as happy when we work eight hours a day as when we worked sixteen. Comfort is not happiness. It may make us better able to enjoy the values which life has to offer but it is not itself a final value. Happiness is found in what we do with our idleness, not in being as comfortable as possible while we are idle. Similarly, while one may enjoy intensely the "glow" of health, if he seeks only this goal he becomes the miserable hypochondriac, mentioned above, who spends his life seeking something which, by itself, is not worth having. And riches, as the life of the miser tells us, may produce only unhappiness. What we should say, therefore, is not that science has made us happier by providing us with a better value scale, but that it has produced instruments by which we *may* attain certain goals, valued either in themselves or as means to more ultimate goals, *provided we wish to do so*. Science has given us the tools but has not told us what to make with them or whether anything we may create with them will increase our happiness one whit. Obviously all that science has done to increase health, comfort, and wealth might equally well be used to intensify our enjoyment of the higher values of life—schools, museums, art galleries, symphonies, moral and political codes, and religions. That our choices have not always been in this direction is not the fault of science; science does not provide the *push* to seek better lives. It provides only the minimal requirements of a good life (since health, money, and play are positive values), but does not tell us whether life *is* good, or whether, if it is *not* good, what kind of life would be better.

3. Science and Scientism

There is, however, another respect in which science may influence our plan for living. In this case it is pure science rather than applied science which seems to determine our value scale. There are two ways in which science may do this—one general, and the other specific. In the general sense, an age of science, such as that in which we are living today, tends to force upon the public its own value scale, which makes truth, wisdom, and learning the highest values of life itself. The result is a philosophy of intellectualism whose manifestation is the giving of the highest esteem to the pursuit of knowledge in all forms. In a society dominated by this point of view the scholar, the teacher, the scientist, even the high-brow so often alluded to in the previous pages, and the man who can answer the sixty-four-thousand-dollar question represent the social elite. Intellectualism becomes the institutionalized approval of reason. Knowing becomes better than enjoying. For such an outlook we do well to appreciate beauty but we do better to know it and understand it; we ought to develop a high type of character and to live a faultless moral life, but we do better to know and understand the principles of right living; we should be commended for being religious, but we should also be able to analyze the religious experience and justify our belief in the existence of God; we may be happy in being a member of a social group, but we ought to understand the structure and organization of society. To *do* a thing may be important, but to *know* what we are doing and why is even better. The logician is therefore preferred to the man who merely thinks, the aesthetician to the artist, the ethicist to the moral man, the theologian to the religious man, and the sociologist to the socially oriented person.

According to this point of view nothing is too sacred or private for study and investigation. The rapid spread of psychology in recent years is largely a result of this attitude. We have psychologists not only of play, of thinking, and of social behavior, but of love, sex, and marriage, of religious conversion, of the aesthetic experience, of imagination and the unconscious, and even of our motives in buying soap and our pleasure in watching television programs. And when wisdom is the highest virtue, ignorance becomes the most complete form of depravity. Science in this role is the Great Emancipator: it is the savior of mankind, not alone from ignorance, authority, and tradition, or even from sin, but from all value experiences which are not properly intensified by a critical examination of what the experiences are and why we enjoy them.

I shall not attempt to evaluate this philosophy of life; it clearly has both merits and demerits. But the position takes on a deceptively attractive form when it defines the supreme value of life not merely as truth, but as *scientific* truth. It frequently becomes so idolatrous of science that it is unwilling even to accept the methods or the results of the behavioral sciences because they are obviously inferior to those of the natural and mathematical sciences. The position has become so strongly advocated in recent years that it has taken on a name: "scientism." [3] It requires that all that poses as knowledge but has not been obtained by the scientific method be abandoned as pseudo knowledge, and made over by applying to its subject matter the principles discussed in the early chapters of this book. Even philosophy must be made scientific—or recognized as nonsense. The humanities must abandon all talk of meanings and values, which are neither observable nor subject to experimentation. Quantitative techniques and instrumental aids must be introduced into all studies. Ultimately, and this has already been attempted, electronic computers will write poems and compose music, and poets and musicians will walk the streets—not because their wares are unappreciated but because they are the final victims of automation.

We need not dwell on this attempt of science to impose its methods on the whole of life. It contradicts the main thesis of this book, and its absurdities are evident. I have argued that science is one of the humanities; the advocates of scientism maintain that the humanities are, or at least should be, one of the sciences. C. P. Snow, as we have already seen,[4] argued that the humanities should not have a monopoly on the word "culture," since there are two kinds of culture—one humanistic and the other scientific. These cultures, he argues, are not antagonistic but supplementary, and the hope of the future lies in increasing the number of men educated in both areas. Scientism would make another proposal: Let us not *share* the word "culture" with the humanists; let us drop it entirely because of its vagueness, because of its emotional overtones, and because of its tendency to import superfluous metaphysical entities on the level of the higher sciences—entities which somehow mysteriously "emerge" from the lower levels.[5] Let us replace "culture" with "scientific humanism." It will then grow rapidly, as science does today, instead of remaining stagnant, as it did under its old name and even as science itself did when in its early years it identified itself with a non-progressing philosophy. It will become productive of human happiness, as science

[3] *Scientism and Values*, ed. H. Schoek and J. W. Wiggins (New York: D. Van Nostrand Co., Inc., 1960); *The New Scientist: Essays on the Methods and Values of Modern Science*, ed. P. C. Obler and H. A. Estrin (Garden City: Anchor Books; Doubleday & Company, Inc., 1962), pp. 31–33; and John Wellmuth, *The Nature and Origins of Scientism*, The Aquinas Lecture, 1944 (Milwaukee: Marquette University Press).

[4] See above, p. 5.

[5] *Readings in Philosophy of Science*, ed. H. Feigl and M. Brodbeck (New York: Appleton-Century-Crofts, Inc., 1953), pp. 8–9. Also see above, p. 200.

has so abundantly done in the past. And it will perhaps share in the financial support which science receives from the foundations and the national government. The analogy which is here employed by the advocates of scientism is so unsufferably bad that I wonder how they can possibly be persuaded by it. Indeed, I suspect at times that they are really arguing with tongue in cheek. Scientism has appeal simply because science is "riding high" today, and I can only reiterate the phrase previously quoted, "When anything calls itself science, beware!" [6]

4. Science and Irrationalism

But, strangely enough, an age of science may also produce a philosophy of irrationalism. For men may rebel when everything is subjected to intellectual scrutiny and nothing remains private. The television viewer who is aroused from his easy chair to answer a telephone inquiry concerning which program he is viewing; the citizen who is stopped on the street by a Gallup pollster and asked how he intends to vote in the next election; and the person who suddenly receives a stroke of good (or even bad) fortune and is condemned by reporters as being "uncooperative" because he refuses to stand for a photograph or to speak into one of the microphones thrust at him—all of these people have grounds for doubting that science and the public *must* have access to all data.

Along with this resentment toward unjustified invasion of privacy there is commonly also a feeling that the enjoyment of one's value experiences is not increased but destroyed when they are subjected to analysis and study. Our most personal values are to be felt and enjoyed, not to be talked about, and certainly not to be classified, correlated, measured, and used as a basis for scientific theorizing. In a well-known poem by Walt Whitman he describes the lecture of an astronomer who explained in a class room, "causing great applause," the movements of the heavenly bodies by means of charts, diagrams, and mathematics. But, said Whitman,

> How soon, unaccountable, I became tired and sick,
> Till rising and gliding out, I wander'd off by myself
> In the mystical moist night-air, and from time to time,
> Look'd up in perfect silence at the stars.

[6] See above, p. 10.

Intellectual scrutiny tears a thing apart and destroys it. Better a simple enjoyment of beauty than a tortured analysis of form and content, and of representation and expression. Better a spontaneous and out-going morality than a critical study of conscience, authority, and the "pleasure principle." Better, even, an ignorance which permits enjoyment than a knowledge which forbids it. Better an appreciation than a criticism; better the heart than the head. This is the philosophy of emotionalism, mysticism, intuitionism, irrationalism—of everything, in fact, that science presumably is *not*.

5. Science and Hidden Values

There is still another, and much more subtle, way in which science is considered to be effective in determining our value outlook. This power of science to produce a value perspective is recognized by many who would accept the arguments which have been raised frequently in the preceding pages, that the non-behavioral sciences have no concern with values but only with existence. They insist that knowledge of what does and does not exist cannot completely determine knowledge of what ought or ought not to exist. Many things, such as pain and poverty, ugliness and war, crime and death, do exist but ought not to; many things, such as health and abundance, beauty and peace, virtue and a happy life, do not exist but ought to. Science can tell us, and does tell us, what exists and how the many things which exist are interrelated. It formulates laws telling us how the objects in the world interact with one another. It explains things in terms of theories about their internal constitution, and in terms of principles, such as gravitation, electrical conduction, chemical combination, growth and evolution, which run through nature. But it does not tell us why there should be such a thing as gravitation, why a world which contains electricity is better than one which does not, why the world is a happier one because of chemical combination, growth, and evolution than it would have been without these factors of change. Science neither finds values in nature nor puts them into nature. Since it discloses cause and effect relations in nature it tells us *which* means will produce *which* ends. But what the ultimate ends are, whether they ought or ought not to be desired, are not the concern of science, and nothing that science can say will help us with such a decision. The final ends are matters of basic desires and preferences, resting, it may well be, on reflective and deliberate processes but not themselves objects which can be characterized truly or falsely. We cannot verify a hope or prove an aspiration.

But those who argue so convincingly for this neutrality of science on human values now turn about and insist that while science does not talk explicitly about values it does concern itself with what is real. Now what is real becomes, by a gradual transition, what is Real (the introduction of the upper-case letter is important). This transition takes us into the area of the metaphysician, whose task is to find out not only what is Real but what is Ultimate Reality. These two are often quite different from one another because what appears to be Real frequently turns out on more careful inspection not to be Real, and we seek a more basic Reality which will both take its place and explain why we were misled into believing that it was Real.

The relevance of this to the problem of how science can determine a philosophy of life can be made clear in terms of an example. The physicist claims that matter (or, in more modern terminology, "mass-energy") is real. If he stops here he remains a physicist. But many scientists do not stop here—at least in their after-hour speculations. Matter becomes not only Real but the Ultimate Reality, and the scientist becomes a Materialist, perhaps even a Dialectical Materialist if he is properly located geographically. Now Materialism, as is well known, argues that matter alone is real. When matter is operated on by forces it brings about all the changes which are exhibited in the events and processes of the world. Thought and mind are simply functions of the brain, which is itself merely a very complex form of organized matter. All values, together with all emotions and feelings, and even all ideas, are simply by-products of matter and material forces. Mind is thus reduced to matter. Such a philosophy, it is claimed, is the only possible one in view of the methods and conclusions of science. Science admits among its facts only those which can be measured; ideas, feelings, and preferences cannot be numbered or weighed, and consequently do not exist as independent realities. Furthermore, as is conclusively shown by injuries to the brain, the use of drugs, and even death itself, consciousness is utterly dependent on the brain for its existence and character, and therefore can have no substance of its own. Thus science itself demonstrates the materialistic philosophy, and any scientist who is at all aware of the broader implications of his method must admit that he is in all logical consistency driven to this position.

But, arrayed against the contemporary Russians and some earlier scientists with similar inclinations, there is an imposing group of scientists, among whom are Eddington, Jeans, Haldane, and Whitehead, who find themselves driven to an idealistic position by the methods and conclusions of science. Eddington,[7] for example, argues that the more we know about matter the less "material" we find it to be; in fact it turns out to be simply a group of shadows—a bundle of pointer-readings. This conclusion, of course, should not surprise us. If we go about the world looking for objects, and carry with us as instruments of detection only measuring devices, we can hardly expect to find anything but pointer readings; if we wear red glasses we shall never find anything but red objects. Science demands, therefore, that there should be a Reality "behind" the measured values; if we measure, there must be something to be measured. All evidence, both inside and outside science, indicates that this Reality is mind. It is that which we discover in the mystic experience, in the appreciation of values, and in the awareness of ourselves as spiritual beings or personalities. What we call "matter" is nothing more than a particular way in which mind manifests itself to us. The ultimate stuff of the world is mind, and matter is simply a deceptive appearance which mind occasionally takes on.

Another example of the way in which science may unconsciously determine one's philosophy of life is to be found in the conclusions of anthropology, sociology, and psychology in the area of human choice. These sciences show that values cannot be in any way embodied in Reality, since they are completely relative. For example, right and wrong are dependent on the society in which we live; what is right for primitive man is wrong in contemporary society, and what is right for Americans is wrong for Hindus. Furthermore, even in a given culture what an individual considers to be right or wrong is largely a matter of education and family training. Hence we must accept relativism as a philosophy of life.

---

7 See the quotation from A. S. Eddington, p. 73.

Presumably this relativity of values to social milieu, education, mass communication techniques, and unconscious yearnings applies throughout our valuational behavior, be it moral, religious, aesthetic, political, economic, or recreational. There are no absolute values, and there is no reason why we should make sacrifices in order to preserve our system of values as against that of any other people. Why should we die for a value which is a function merely of geography, or of the historical age in which we happen to live, or even of the way in which our glands happen to be functioning at the moment? "Oughtness" is a mere phantasy and has no foundation in Reality. No desire, however intense, can guarantee its own satisfaction. Jeremy Bentham said that science ought not to use the word "ought" except for the purpose of saying that it ought not to use it.

But certain further conclusions from the behavioral sciences lead to an even more striking philosophy of life. We must all become Determinists. Recent studies in motivational research and mass communication have shown that many of our value preferences are not "free choices" at all. We buy a certain detergent not because we "want" it but because we are attracted to the red package which contains it, or because we have had its name dinned into our ears repeatedly by the commercials on radio and television programs, or even because something remotely connected with it satisfies an infantile desire which we have suppressed for many years. We have all learned that our political preferences can be determined through propaganda techniques. Science, therefore, plays an important role in determining our philosophy: it is the Great Destroyer of Illusions, the Great Exploder of Fond Hopes. It tells us that the choices which we make in selecting our goals are not really free choices, and that the goals which we presumably select are not inherently worthwhile. We are the victims of forces over which we have no control, and are no more able to plan our futures than is the stone which is crashing down the mountainside. Any earnest pursuit of values is therefore futile and can only end in defeat and frustration. We shall all do well to face this bitter truth and not be carried along by empty illusions.

But now suppose we *do* face this truth. What shall we *then* do? Shall we stop making all choices? If we do, we make a choice, namely, the choice not to make any choices. And if we do not make any choices we shall, of course, do nothing. Now this will work very well so long as we are physically comfortable and not too hot or too cold, not in pain, not too tired, not worried, not apprehensive about the future, not seeing anything ugly or hearing discordant tones, and so on through the list of negative values which normally spur us on to activity. If we persist in refusing to make any choices we shall soon die. The fact that we do not remain stubborn in this belief in the ineffectiveness of choices indicates that we do not really accept it. Sooner or later we do make a choice, and we act on the basis of this decision. We get hungry and decide to seek food even though we know that the desire for food is merely a function of our glands and is "not embodied in Reality." Even in this case we *could* decide not to eat; rats can be conditioned to starve themselves, and men have died simply because they refused to eat. We may, of course, decide to act on other grounds. We may act merely because *not acting* makes us uncomfortable. Possibly we begin to feel guilty because we are so indolent. Perhaps we become unhappy because we are not "doing what comes naturally."

The point of all this is that whatever science may tell us about the relativity of all choices and our inability to make free choices, we shall nevertheless continue to make choices. For to be human is to make choices. To decide not to make choices is, in fact, a choice: it is the choice not to be human. We can make this choice but few of us ever do. We prefer to remain human.

The argument which we have been examining is indeed a very strange one. It says that our studies of human nature have shown that there are no free choices and all choices are purely relative, the conclusion then being that we should give up all choices. Thus by studying human nature we must conclude that we cannot be human. No doubt all of us, having seen examples of human depravity, have at times questioned the desirability of allowing mankind to propagate and continue to exist on the earth. But this is a rare attitude and probably none of us would try to argue convincingly for it. In fact if we used human depravity as an argument for exterminating the human race we should be implicitly admitting that men *do* make choices (bad ones), and that these choices *are* often powerful factors for evil. Hence there *are* free choices and some of these choices are *objectively* bad.

To claim that the behavioral sciences—particularly anthropology, sociology, and psychology—determine our philosophy of life by exposing the emptiness of all values and the absurdity of free choice is to look upon these sciences only at the lowest of the three levels of awareness. As a matter of fact what they disclose, when critically understood, is the omnipresence of a dynamism of values in human life. It is no accident that preferences are called "social forces." Indeed, in the case of the reformer the "force" which compels him to act may be quite *unsocial,* since it is a value which may not be approved by his particular group. Men do choose to die for their country, their family, their God, and truth. To say that these values are not "embodied in Reality" is to identify Reality with a physical world from which values have been previously excluded. There certainly is a Reality which man experiences when, as a human being, he feels the drive of moral, religious, aesthetic, and intellectual values. This is where philosophy makes up for the limitations of science. If we are to base our design for living on the sciences we must include *all* the sciences—not only anthropology, sociology, and psychology, but the normative inquiries such as ethics, aesthetics, philosophy of religion, and logic as well. These studies (even though some are called "philosophies" and "meta-sciences" rather than "sciences") make perfectly clear that there are objective values—values which are strong or weak, high in quality or low in quality, pure or mixed, instrumental or ultimate. "Oughtness" does have status in the world, and if we deny this we are not facing the facts. A story is told of the English philosopher, G. E. Moore, who had written a paper to be presented before the British Academy. He had worked long and hard on the address, but was not able to revise it to his satisfaction. On the day of the lecture, as he was leaving home, he expressed his concern to his wife. She replied, comfortingly, "Don't worry about it; they'll enjoy it." "Well," said Moore, "if they *do* they'll be *wrong.*" Moore obviously believed that it was objectively wrong to like something which was inferior in quality.

We have now seen two examples of the way in which science may determine one's philosophy of life: it may cause him to be a Materialist or an Idealist, and it may cause him to see all value pursuits as fruitless and empty or as filling his life with meaning and purpose. Neither of these positions has been developed in more than outline form. But perhaps enough has been said to indicate that we *are* unable to infer a design for living on the basis of science alone. We must start with certain *facts* about science and proceed *by correct logic* to our conclusion. And if our conclusions contradict one another, as seems to be the case, presumably we have either started with different facts or used bad logic. This can mean to the man in the street, who must take both the scientific facts and the logic of our argument more or less on faith, only that science probably does not have direct implications for *any* world view. This is certainly in accord with the view developed in Chapter 12. There we saw that science is distinguished from philosophy by being concerned with the specific rather than the general, by studying the part rather than the whole, and by investigating nature and our preferential responses rather than our deeper evaluations of nature. The question is therefore whether from knowledge of the parts, of the specific aspects, and of the "surface" of nature, we can infer anything about the whole, the general aspects, and our more profound understanding of nature.

Certainly if there are any such inferences they will be tenuous and loose, and the only conclusion we can draw is that any philosophy of life, whether based on science or not, must be precarious and uncertain. We cannot demonstrate a philosophy of life in the same way that we can demonstrate a theorem in mathematics. We cannot even show one design for living to be better than another in the same way that we can show the theory of evolution to be a better explanation of life than the theory of fixed species. General facts, the universe as a whole, and our evaluation responses are hidden and elusive. We cannot become aware of them through our senses, we cannot measure them, we cannot experiment with them, and we cannot even classify or correlate them with any high degree of success. We cannot successfully arrange them in formal, deductive systems. We cannot therefore handle them by the usual techniques of science. All that we can do is hunt for them, talk about them, think about them, and speculate about them with such rational techniques as we have at our command. In many respects the best way to get these facts is *through* science, but only in this sense can we say that science determines our philosophy of life.

6. Science and the Design for Living

The philosophy of life which has been defended in the preceding pages (insofar as *any* such philosophy has been presented) lies at some point between a rational and an irrational one. We have argued that the problem of effective living is that of making wise choices. Since our choices lie among the ultimate values of life—truth, beauty, goodness, piety, health, wealth—science cannot help us greatly. Science cannot tell us whether we shall find the life of the scholar more satisfying than the life of the artist or of the doctor, nor whether we shall find the life of the businessman more rewarding than the life of the social worker or of the religious teacher. Here "reasons of the heart," together with accidents of birth, environmental circumstances, native capacities and endowments largely make our choices for us. To be sure, we can "think about" these alternatives, endeavor to inform ourselves through history and literature concerning what they have to offer, explore them in imagination, and finally arrive at our decision. But to say that such a decision is irrational is perhaps a little misleading, just as it would be misleading to say that in science we *never* make decisions of this kind. When we are setting up a value scale we know that we *must* use our feelings even when we try to be objective about the whole matter, and in science we know that we *do* use our feelings but are convinced that this can probably be avoided if we try hard enough. In this sense both philosophy and science are at once rational and irrational: philosophy tries to be rational but acknowledges that many of its decisions are, and perhaps must be, irrational; science tries to be rational and usually succeeds, though the pressure of the irrational is often very great.[8] In this sense the philosophy of life of a scientific society attempts to bring together intellect and emotion.

This compromise position does not accept the extreme intellectualism which argues that knowing and understanding are always better than "mere feeling" and should replace it. Knowledge and zeal are not in inverse proportion to one another. Nor does the view accept the extreme irrationalism which argues that

[8] "If the Baconian method of observation should bring complete objectivity, that objectivity would . . . come as near to idiocy as perfect subjectivity would to madness." A. L. Porterfield, *Creative Factors in Scientific Research* (Durham: Duke University Press, 1941), p. 122.

our value experiences are private and not subject to examination by reason. If human living is the making of choices, and if happy living is the making of wise choices, then our value enjoyment can often be enhanced only if we think about the choices we are making. This conclusion has been shown both by the fact that as we grow from youth to maturity we *do* think more about our choices, and by the fact that when we think about our values we *do* often discover that our value experiences are widened, deepened, and rendered more satisfying. Many people, of course, pass through life preferring not to think, and even, perhaps, never thinking. Poincaré said that there are two ways of getting through life easily—one way is to believe everything, the other is to doubt everything; both ways save us from thinking. On the other hand, many people choose always to think about everything, and thus run the risk of destroying their value enjoyments. But the man who never thinks and the man who always thinks are extremely rare, and usually succeed only in being unhappy as a result of their efforts. The wise man knows when to think and when not to think.

Such a design for living does, however, depend on applied science in a very important way, and is therefore more and more effective just to the extent to which the pure science on which it is based is also well grounded and reliable. Having selected our ends (which neither pure nor applied science can do for us, but in which philosophy in its role as the evaluator of values can influence our choice) we must search for the instruments by which these ends may be attained. Now we have seen that when an effect is something desirable, the relation of cause-to-effect becomes the relation of means-to-end. If A is the cause of B, then if we want B we seek out A as the means by which it can be produced. An effect may have many causes, in which case if the effect is desired there will be many ways of satisfying it. A causal law may be *necessary*, and the desired effect can then be *certainly* attained, provided the means are under our control. Or a causal law may be *statistical*, and the desired effect can then be only *probably* attained, again provided the means are under our control. Cause and effect laws of these kinds are discovered by applied scientists, and a science which has established a large number of such laws is a mature and successful one. But they must be "put to work" by the individual himself in his role as artisan or worker. Consequently if he wishes to become rich, applied economics can tell him how to do so; if he wishes better government, the applied science of politics will come to his aid; if he wants better health, physiology and medicine will provide him with the necessary information; if he wishes to be more religious, the applied science of ministerial training and proselytizing will be of value to him; and if he wants greater physical comfort, applied physics, chemistry, and biology will provide the necessary information. In all cases where the problems are complicated and the decisions difficult to make the use of the applied behavioral science will require appeal not only to the corresponding pure sciences but also to the philosophical, humanistic, and meta-scientific correlates of these sciences.

But to live effectively requires more than the philosophical knowledge which enables us to make wise choices among conflicting values, and more than the pure and practical knowledge provided by the sciences and those aspects of "applied philosophy" which tell us how to select our means to the attainment of our preferred goals. It depends on the skillful application of this information to the problems of living. To live happily is more than to know how to live happily, just as to build bridges is more than to know how to build bridges. Skill is required in the use of means to attain ends. This is acquired only in the business of daily living—a kind of trial and error process by which we learn to make highly discriminatory judgments, to proceed often on hunch, to rely on tact, and through the school of experience to learn how to fit our ideal pattern of values to the complex world in which we are obliged to live.

Lundberg asks, "Can science save us?" Then he answers,

> Yes, but we must not expect physical science to solve social problems. We cannot expect penicillin to solve the employer-employee struggle, nor can we expect better electric lamps to illumine darkened intellects and emotions. We cannot expect atomic fission to reveal the nature of the social atom and the manner of its control. If we want results in improved human relations we must direct our research to the solution of these problems.

In reply to those who are still skeptical and unimpressed with the promise of social science, he asks, "If we do not place our faith in social science, to what shall we look for social salvation?"[9]

[9] G. A. Lundberg, *Can Science Save Us?* (New York: Longmans, Green & Company, 1947), p. 104. Courtesy of David McKay Company, Inc.

My reply is twofold: In the first place, social (behavioral) science offers great promise. But the term "social science," as we have seen, is used in a very confused manner: it may mean "pure behavioral science," "applied behavioral science," "the philosophy or meta-science of social behavior," and "social work." If the term includes all of these studies and activities, and does not confuse them one with another, social science can contribute greatly to the solution of many of our current problems. In the second place, many problems of effective human living lie as much in the individual as in the group of which he is a member. In choosing a design for living it is the *person* who makes the final decision, be it wise or unwise. And since this decision lies among the ultimate and final values which life has to offer, philosophy, in its attempt to integrate nature, life, and value, must play a dominant role. Lundberg is right in objecting to the sharp conflict between the natural sciences, which have no concern with values, and the humanities, which are value-ridden; such a dichotomy only fosters antagonisms and misunderstandings. The behavioral sciences do, in a way, bridge this gap, for they are concerned with values. But if the behavioral sciences are not to be identified with the natural sciences, neither are they indistinguishable from the humanities. They do not examine the deeper, more ultimate values and the problems which arise with regard to presuppositions and interrelationships. Their disregard for these aspects of behavior is not to condemn them; it is merely to recognize the need for division of labor. The humanist has been assigned this task, and he is performing it to the best of his ability. But even he cannot complete his assignment without knowing and using the results of the natural and behavioral sciences. Thus any partitioning of the intellectual process into water-tight compartments is fatal to the development of that type of understanding which we associate with the truly cultured man.

Happy living, then, depends on three things. (1) Wise choice as to ends. This choice may be ultimately irrational, but it can be "guided" by reason. When the philosopher tells us about the general features and pattern of the universe, and about the nature of the value experiences, he is trying to help us in making this basic choice. (2) Wise selection of the means to these ends. This is based on information concerning cause and effect relations holding in the world and in our experience—information provided mainly by the scientist. (3) Effective use of skills in the employment of means to these ends. This is learned from practice, or from the practitioners — the businessman, the politician, the doctor, the engineer, the religious and moral teacher, the art trainer, and the practical logician. The role of science in this design for living is an important — we might even say a dominant and essential — one. But the world will be neither saved nor destroyed, barring accidents, by science alone. It will be saved by men of knowledge and vision who use science to promote the good life; it will be destroyed by men of ignorance and myopia who use science to promote the bad life. In the former case science will be praised, and in the latter case it will be blamed; but it will deserve neither. For science is not a slave and not a master; it does not exist wholly to serve nor wholly to lead. It is slave to only one master — truth; but it is master, for the truth shall make us free.

Name Index

Subject Index